培生科学虫双语百科
奇妙物理

Changing Shape

变化的形状

英国培生教育出版集团 著·绘
徐 昂 译

电子工业出版社
Publishing House of Electronics Industry
北京·BEIJING

Original edition, entitled SCIENCE BUG and the title Changing Shape Topic Book, by Tanya Shields published by Pearson Education Limited © Pearson Education Limited 2018
ISBN: 9780435195397

All rights reserved. No part of this book may be reproduced or transmitted in any form or by any means, electronic or mechanical, including photocopying, recording or by any information storage retrieval system, without permission from Pearson Education Limited.

This adaptation of SCIENCE BUG is published by arrangement with Pearson Education Limited.
Chinese Simplified Characters and English language (Bi-lingual form) edition published by PUBLISHING HOUSE OF ELECTRONICS INDUSTRY, Copyright © 2023.
For sale and distribution in the mainland of China exclusively (except Hong Kong SAR, Macau SAR and Taiwan).

本书中英双语版由Pearson Education（培生教育出版集团）授权电子工业出版社在中华人民共和国境内（不包括香港、澳门特别行政区及台湾地区）独家出版发行。未经出版者书面许可，不得以任何方式抄袭、复制或节录本书中的任何部分。

本套书封底贴有 Pearson Education（培生教育出版集团）激光防伪标签，无标签者不得销售。

版权贸易合同登记号　图字：01-2022-2381

图书在版编目（CIP）数据

培生科学虫双语百科. 奇妙物理. 变化的形状：英汉对照 / 英国培生教育出版集团著、绘；徐昂译. --北京：电子工业出版社，2024.1
ISBN 978-7-121-45132-4

Ⅰ.①培⋯ Ⅱ.①英⋯ ②徐⋯ Ⅲ.①科学知识-少儿读物-英、汉 ②物质-形态-变化-少儿读物-英、汉 Ⅳ.①Z228.1 ②O4-49

中国国家版本馆CIP数据核字（2023）第035074号

责任编辑：李黎明　文字编辑：王佳宇
印　　刷：河北迅捷佳彩印刷有限公司
装　　订：河北迅捷佳彩印刷有限公司
出版发行：电子工业出版社
　　　　　北京市海淀区万寿路173信箱　邮编：100036
开　　本：787×1092　1/16　印张：35　字数：840千字
版　　次：2024年1月第1版
印　　次：2024年2月第2次印刷
定　　价：199.00元（全9册）

凡所购买电子工业出版社图书有缺损问题，请向购买书店调换。若书店售缺，请与本社发行部联系，联系及邮购电话：（010）88254888，88258888。
质量投诉请发邮件至zlts@phei.com.cn，盗版侵权举报请发邮件至dbqq@phei.com.cn。
本书咨询联系方式：010-88254417，lilm@phei.com.cn。

使用说明

欢迎来到少年智双语馆！《培生科学虫双语百科》是一套知识全面、妙趣横生的儿童科普丛书，由英国培生教育出版集团组织英国中小学科学教师和教研专家团队编写，根据英国国家课程标准精心设计，可准确对标国内义务教育科学课程标准（2022年版）。丛书涉及物理、化学、生物、地理等学科，主要面向小学1~6年级，能够点燃孩子对科学知识和大千世界的好奇心，激发孩子丰富的想象力。

本书主要内容是小学阶段孩子需要掌握的物理知识，含9个分册，每个分册围绕一个主题进行讲解和练习。每个分册分为三章。第一章是"科学虫趣味课堂"，这一章将为孩子介绍科学知识，培养科学技能，不仅包含单词表、问题和反思模块，还收录了多种有趣、易操作的科学实验和动手活动，有利于培养孩子的科学思维。第二章是"科学虫大闯关"，这一章是根据第一章的知识点设置的学习任务和拓展练习，能够帮助孩子及时巩固知识点，准确评估自己对知识的掌握程度。第三章是"科学词汇加油站"，这一章将全书涉及的重点科学词汇进行了梳理和总结，方便孩子理解和记忆科学词汇。

2024年，《培生科学虫双语百科》系列双语版由我社首次引进出版。为了帮助青少年读者进行高效的独立阅读，并方便家长进行阅读指导或亲子共读，我们为本书设置了以下内容。

（1）每个分册第一部分的英语原文（奇数页）后均配有对应的译文（偶数页），跨页部分除外。读者既可以进行汉英对照阅读，也可以进行单语种独立阅读。问题前面的 🖉 符号表示该问题可在第二部分预留的位置作答。

（2）每个分册第二部分的电子版译文可在目录页扫码获取。

（3）本书还配有英音朗读音频和科学活动双语视频，也可在目录页扫码获取。

最后，祝愿每位读者都能够享受双语阅读，在汲取科学知识的同时，看见更大的世界，成为更好的自己！

电子工业出版社青少年教育分社
2024年1月

Contents 目录

Part 1　科学虫趣味课堂　　　　　　　　　　　　　　／1

Part 2　科学虫大闯关　　　　　　　　　　　　　　　／31

Part 3　科学词汇加油站　　　　　　　　　　　　　　／47

Part2译文

配套音视频

Changing Shape

Some materials change shape easily.

Some materials do not seem to change shape at all.

Word Box
bend
squash
stretch
twist

We can change the shape of materials if we **bend**, **twist**, **squash** or **stretch** them.

Think about the materials you have touched today.

1. Which materials change shape easily?
2. Which ones do not?
3. Which other everyday objects change shape easily?

Modelling clay can change shape easily.

变化的形状

一些材料的形状容易被改变。
一些材料的形状似乎完全改变不了。

单词表

bend 弯曲
squash 挤压
stretch 拉伸
twist 扭转；捻

我们可以通过**弯曲**、**扭转**、**挤压**或**拉伸**来改变材料的形状。

思考一下你今天触摸过的材料。
1 哪些材料的形状容易被改变？
2 哪些材料的形状不易被改变？

橡皮泥可以很容易地被改变形状。

3 有哪些日常物体的形状容易被改变？

Stretching

Look at the pictures.

1 In what ways are they alike? Think about what is happening and what materials have been used to make the things in the pictures.

People use elastic bungee cords to jump off high buildings and bridges.

If you blow up a rubber balloon too much it will pop!

Gymnasts have clothing that stretches so they can move freely.

拉伸

观察图片。

1 这些图片有什么相似之处？思考一下发生了什么，同时思考一下图片中的活动器材是用哪些材料做的？

人们系上有弹性的蹦极绳，从高耸的建筑物或桥上跳下去。

如果你把一个气球吹得太大，它就会爆开！

体操运动员所穿的服装弹性十足，这样他们才能轻松地做出各种动作。

Plastic wrap is used to stop food drying out.

The biggest elastic band ball weighs the same as two rhinos (4,097 kg)!

All of these objects are designed to stretch.

2 What would happen if these materials did not stretch?

3 What materials would not be very good for making a bungee cord, plastic food wrap, or a balloon?

Activity

Try stretching some elastic bands. Be careful not to stretch them too much or they might snap!

1 What does it feel like?
2 Do all elastic bands feel the same when you stretch them?

封上保鲜膜可以防止食物的水分流失。

最大的橡皮筋球和两只犀牛一样重（4,097千克）！

这些物体都是可以拉伸的。

2 如果这些材料不能拉伸会发生什么？
3 哪些材料不适合用来制作蹦极绳、保鲜膜或者气球？

活动

试着拉伸一些弹力带，但不要用力过猛，否则它们可能会断掉！
1 拉伸弹力带是什么感觉？
2 当拉伸不同的弹力带时，感觉一样吗？

Rubber

Rubber is an amazing material! You can squash it, bend it, stretch it, twist it and **bounce** it. But where does it come from? Believe it or not, rubber comes from trees!

Word Box
bounce
latex
rubber

The white liquid from rubber trees was used to make shoes.

Rubber was first discovered hundreds of years ago by people living in Central and South America. They collected the white liquid that oozed from the rubber trees and smeared it onto their feet to make shoes.

The white liquid collected from rubber trees is called **latex**.

橡胶

橡胶是一种神奇的材料！你可以将橡胶挤压、弯曲、拉伸、扭转和**反弹**。那橡胶是从哪里来的呢？信不信由你，它的来源是树！

> **单词表**
> bounce 反弹
> latex 乳胶
> rubber 橡胶

从橡胶树中流出的白色液体可以用来制作鞋子。

橡胶是由几百年前生活在中南美洲的人首次发现的，他们将橡胶树渗出的白色液体收集起来，涂抹到脚上，制成鞋子。

从橡胶树收集而来的白色液体叫作**乳胶**。

Did you know?

Rubber trees are difficult to find, so scientists have developed a way to make rubber in their laboratories. We use the same method as scientists to create everyday things from rubber.

Rubber safety tiles are used in playgrounds.

Rubber is used to make tyres for cars.

Rubber is still used today to make thousands of different things.

How many things can you think of?

你知道吗？

橡胶树很难被发现，因此，科学家们总结了一些方法在实验室里生产橡胶。我们现在也采用与科学家们一样的方法用橡胶生产各种各样的日常生活用品。

由橡胶制成的安全地砖被用在操场上。

橡胶被用来制造汽车轮胎。

橡胶在今天仍然被应用于生产成千上万种产品。

你能想出多少种由橡胶制成的东西？

Bending Metal

Have a look at the **metals** around you. What do you see? Do you think you could bend them easily? Metal is usually very strong and not very bendy.

Word Box
metal

There are a few metals that are very bendy. We can use these metals to make artwork and jewellery, or simply to cover food.

Humans have been making metal jewellery for over eight thousand years. This bracelet was made in Egypt over 2,000 years ago.

We can easily bend kitchen foil because it is very thin.

Sometimes we need to shape metal that is not easy to bend. A blacksmith uses heat to help bend metals. This is a dangerous job and the blacksmith has to wear thick gloves and clothes to protect them from the heat. Once the metal is hot enough the blacksmith can bend the metal into shape.

弯曲的金属

观察你身边的**金属**，你观察到了什么？你认为自己可以轻易地让金属弯曲吗？金属通常较为坚硬且不易弯曲。

单词表
metal 金属

有少数金属的弯曲性很好，我们可以用这些金属制作艺术品和珠宝，也可用来简易地包装食物。

人们制作金属珠宝已经有八千多年的历史了，这个手镯诞生于2000多年前的埃及。

我们可以轻易地卷曲锡箔纸，因为它很薄。

有时我们需要改变不易弯曲的金属的形状。铁匠将金属加热，这样就可以将其弯曲。这是很危险的，所以铁匠需要戴上厚厚的手套，穿上厚实的衣服来保护自己不被烫伤。一旦金属达到一定温度，铁匠就能将金属弯曲并打磨成想要的形状。

As the metal gets very hot it changes colour.

Artists also use heat to bend metals. Once the metal cools it stops being bendy.

Make your own

Have you seen any bendy metal art?
Twist and bend some kitchen foil to make your own piece of art.

Imagine if a bridge was made from a bendy material.
What would happen when cars and lorries drive over it?

当金属被加热到一定温度时,它的颜色会发生改变。

艺术家也会利用加热使金属弯曲。一旦金属冷却,它就不再容易弯曲了。

自己动手做

你看到过弯曲的金属艺术品吗?
扭转或卷曲锡箔纸,制作一个自己的艺术品。

想象一下,如果一座桥是由可弯曲的材料制成的。
当汽车和卡车在桥上行驶时会发生什么?

Science Skills

Bendy Ruler – Predict it!

Rulers can be made from different materials and can be different sizes.

1. Find some different rulers, for example, wooden, plastic shatterproof and metal.
2. Look at the rulers. Which one do you think is the most bendy?
3. Put the rulers in order from those you think are most bendy to least bendy.
4. Test the rulers. Were you right?

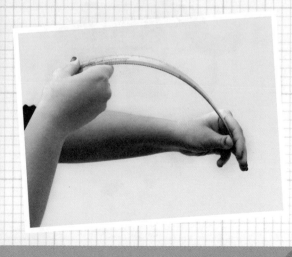

科学技能

弯曲的直尺——预测一下吧!

直尺可由不同的材料制作成不同的尺寸。

1. 找到一些不同材料的直尺,例如,木质的、抗震塑料的和金属的。
2. 观察这些直尺,你认为哪把直尺最易弯曲?
3. 从你的角度,将直尺按照从最易弯曲到最不易弯曲的顺序排列。
4. 用这些直尺测试一下,你的猜测是正确的吗?

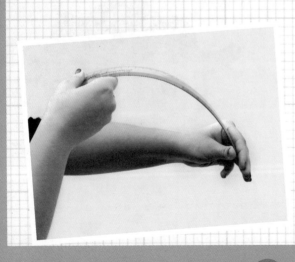

Rope Twisting

Twisting is just one way you can change the shape of materials. **Rope** is made by twisting long, stringy materials together, and can be made from **plant fibres**.

The Ancient Egyptians made special tools to help them make rope. These tools held one end of the rope so that the rope makers could use both hands to twist.

Word Box
plant fibre
rope

The first ever ropes were made by twisting vines (long plant fibres) together.

Egyptians used special tools to make rope from plant fibres.

1a Why do you think we twist threads together to make rope?

1b Why don't we just use single threads?

You will need: some string, some scissors...
Cut 3 pieces of string all 30 cm long. Twist the string together to make your own rope.

缠绕绳子

缠绕是一种能改变材料形状的方法。**绳子**是通过将长的纤维材料缠绕在一起制成的，绳子也能由**植物纤维**制成。

单词表
plant fibre 植物纤维
rope 绳子

古埃及人发明了特殊的工具来帮助他们制作绳子，这些工具能固定绳子的一端，这样生产者就能用双手将绳子的细线缠绕到一起了。

最初的绳子是由拧在一起的藤蔓（长的植物纤维）制成的。

埃及人曾经借助特殊的工具将植物纤维制成绳子。

1a 你认为我们为什么要把线缠绕在一起来制作绳子？
1b 为什么我们不直接用一根线？

你将需要：一些细线、几把剪刀……
剪3根30厘米长的细线，把3根细线缠绕在一起，制作你自己的绳子。

Making Yarn and Weaving Cotton

Cotton comes from a plant.

Cotton spinners use twisting to make **yarn** from cotton. Twisting strands of yarn together makes it stronger. The yarn can then be used to make fabric which can then be made into clothes.

Word Box
cotton
yarn

This is how cotton spinners work.

First they remove all the seeds, this leaves behind soft cotton wool balls. Then they pull and stretch the cotton into long bundles.

They use a tool called a spinner to twist the cotton and make yarn. People weave yarn together to make material.

纺纱织棉

棉花的来源是植物。

纺织者利用缠绕的手法用棉来纺**纱**。将纱线缠到一起可以增大它的韧度，然后就可以用纱来制作布料，从而制作衣服。

> **单词表**
> cotton 棉花
> yarn 纱线

纺织者的工作流程就是这样的。

首先，他们去除棉花的所有种子，剩下柔软的棉花球。然后经过拉、押，他们将棉花球变成长长的棉花团。

他们用纺纱机将棉花捻到一起来制纱，人们一起纺纱来制作新的材料。

One set of threads hangs from top to bottom, these are called the warp threads. Another thread is then passed under, then over each of the hanging threads. These are called the weft threads. The result of weaving threads is a fabric.

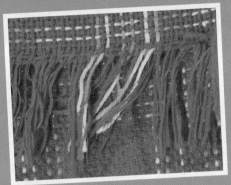

You can also knit yarn together to make clothes. Knitting bends and knots the yarn together.

Activity

Use a hand lens or digital microscope to look closely at your clothes.

Look at how the threads are twisted together.

1a How do you think your clothes were made?

1b Were they knitted or woven?

Weaving - you try it!

You can make paper mats by weaving strips of paper.

从上到下悬挂着的一组纱线叫作经纱;从左到右横着不间断地上下穿过经纱的纱线叫作纬纱。纺织纱线就能得到布料。

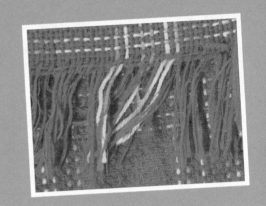

你也能将纱编织到一起来制作衣服。编织能让纱弯曲并打结。

活动

用放大镜或电子显微镜来仔细观察你的衣服。

观察细线缠绕的方式。

1a 你认为你的衣服是如何制作的?

1b 你的衣服是编织的还是纺织的?

纺织——动手试一试!

📄 你可以尝试着用纸条编织一个小纸垫。

Making Balloons

The first modern balloon was made in 1931 and was shaped like a cat's head.

Word Box
mould

Today balloons are made in factories, and they come in lots of different shapes and colours. They are made from a stretchy white liquid called latex.

Yellow latex has been used to cover these moulds.

Balloon makers mix dye into the latex, to create colourful balloons.

The balloon **moulds** are then dipped in the liquid latex. A thin layer of latex sticks to the mould.

The latex-covered moulds are then heated and washed.

制作气球

第一个现代气球生产于1931年，它的形状像一个猫头。

如今的气球是在工厂里生产出来的，有不同的形状和颜色。气球是由乳胶这种有弹性的白色液体制成的。

单词表
mould 模具

黄色的乳胶被用来覆盖在模具上。

气球制造商将染料放进乳胶中，这样可以生产出五颜六色的气球。

然后，将气球**模具**浸入到液体乳胶中，一层薄薄的乳胶附着在模具上。

之后，将附有乳胶的模具加热、清洗。

Finally, the balloons are removed from the moulds and checked. A machine is used to blow up the balloons to make sure they don't pop too easily.

Modelling balloons

You will need: some long, thin modelling balloons, a balloon pump...

1. Pump air into some balloons then twist, squash and bend them.
2. What different shapes can you make?
3. Can you make a balloon animal? You might need to use more than one balloon.

最后，从模具上取下气球，检查是否完好。再用机器向气球里吹气，确保它不会轻易地爆开。

模型气球

你将需要：一些长长的、细细的气球、打气筒……

1 给气球打气，然后将气球扭转、挤压或弯曲。

2 你能制作出多少种形状？

3 你能制作一个动物气球吗？你可能需要用到不止一个气球。

Making Models

You can use dough to make models, decorations, even candle holders. When the dough is **soft** it can be shaped easily.

Word Box
harden
soft

Tips for creating eye-catching creatures

Here are some ideas for how to change the shape of dough to make different animals.

Dough has been stretched and squashed to make model animals.

Roll and squash to make a long, thin shape like a snake.

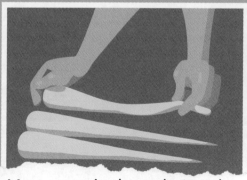

You can pinch and stretch small bits of dough to make hedgehog-like spikes.

制作模型

你可以用面团制作模型、装饰物，甚至是烛台。当面团柔软时，它能轻易地被改变成各种形状。

> **单词表**
>
> harden 变硬
> soft 柔软的

小提示：如何制作引人注目的生物

这里有一些关于如何改变面团的形状来制作不同动物的想法。

面团可以被拉伸、挤压，这样可以被制成动物模型。

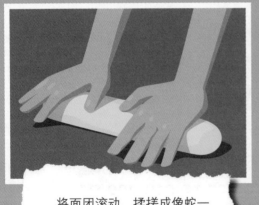

将面团滚动、揉搓成像蛇一样的，长而细的形状。

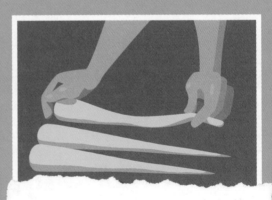

你可以揉捏小块的面团，这样可以制作出和刺猬的刺一样的形状。

Bend and roll to make shell-like spirals, and you will soon make a snail.

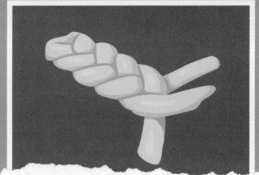

Twist and plait to make rope-like patterns like you might find on a horse's tail.

If you leave the dough to dry out it will eventually **harden**, and you will no longer be able to change its shape.

Potters create plates, bowls and mugs in the same way. They mould soft clay into their chosen shapes before leaving them to dry out and harden, or baking them in an oven.

Make some modelling dough from flour and water

1 What steps do you need to take to make the dough?

2 Make a model out of your modelling dough. Leave it in a warm place to dry out. How does your model change?

 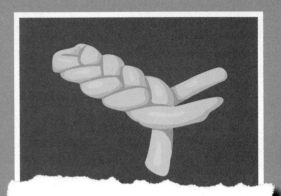

将面团卷曲、回旋成贝壳状的螺旋，然后你就能很快地制作出蜗牛。

将面团扭曲并编成辫子一样的形状，这看起来像一条马尾。

如果你让面团自然风干，它最终会**变硬**，之后你就不能再改变它的形状了。

制陶者用同样的方式制作盘子、碗和杯子等。在黏土风干、变硬之前，他们将软黏土放进自己选好的模具里，或者将它们放进炉子里烘烤。

用面粉和水制作一些模型面团

1. 你需要采取哪些步骤才能制作出面团？
2. 用你的模型面团制作一个模型，把模型放到温暖的地方等它变干。你的模型会发生什么变化？

Changing Shape

Think about the materials you have touched today.

1   Which materials change shape easily?

   _____

2   Which materials do not change shape easily?

   _____

3   Work in a pair to finish these sentences. You can use words or pictures to help you. The first sentence has been done for you.

You can change the shape of a *ruler* by *bending* it.

You can change the shape of a _____ by _____ it.

You can change the shape of a _____ by _____ it.

You can change the shape of a _____ by _____ it.

You can change the shape of a _____ by _____ it.

Change the Shape

Look at the diagrams. Then draw a circle around the materials that you think will change shape easily.

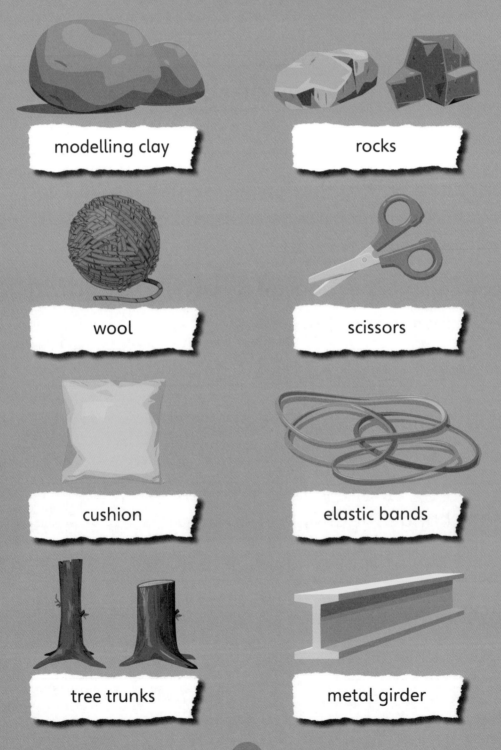

modelling clay

rocks

wool

scissors

cushion

elastic bands

tree trunks

metal girder

How Was It Changed?

Look at the items below.

1. Think about how their shape has been changed.

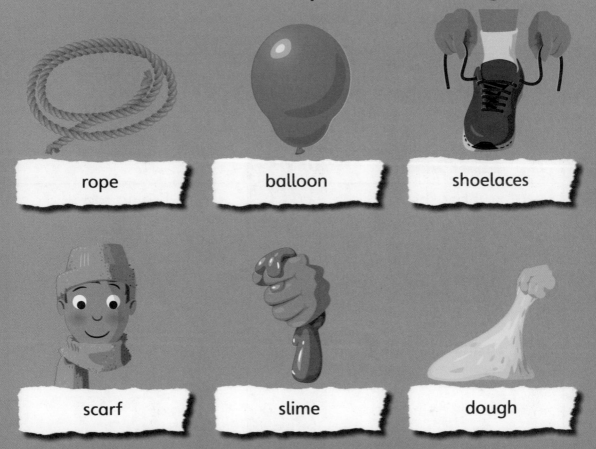

2. Sort the items by how they have changed shape in the table below.

Stretched	Bent	Squashed	Twisted

Changing Materials to Make Things

Match each material to something that is made from it. The first one has been done for you.

Material	Use
elastic 	knitted jumper 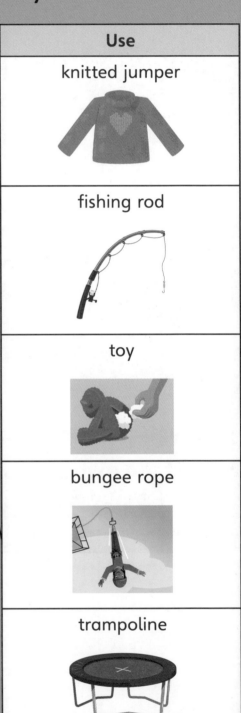
rubber sheet	fishing rod
strand of wool	toy
thin piece of wood	bungee rope
stuffing	trampoline

Terrible Trampolines

A trampoline uses lots of stretchy materials.

1. What would be the worst material you could use to make a trampoline?

   _____

2. Why is your choice such a poor material for a trampoline?

   _____
   _____

3. What materials would not be very good for making a bungee cord, or a balloon?

   _____
   _____

Science Skills

Stretchy Putty Races - Graph it!

Isla and Vash have made their own stretchy putty. They want to find out whose putty stretches fastest.

They made sure they both had the same amount of putty.

They marked out a starting line at the top of the window.

They each placed their putty on the start line.

They recorded their results every two minutes.

Here are the results from their putty race:

	Time from start (minutes)							
	2	4	6	8	10	12	14	16
Distance stretched Isla (cm)	1	2	4	6	7	8	10	12
Distance stretched Vash (cm)	1	1.5	3	4	5	6	8	10

1 Whose putty won the race?

   _____

2 How far did the winning putty stretch?

   _____

3 Finish the graph to show how quickly the winning putty stretched.

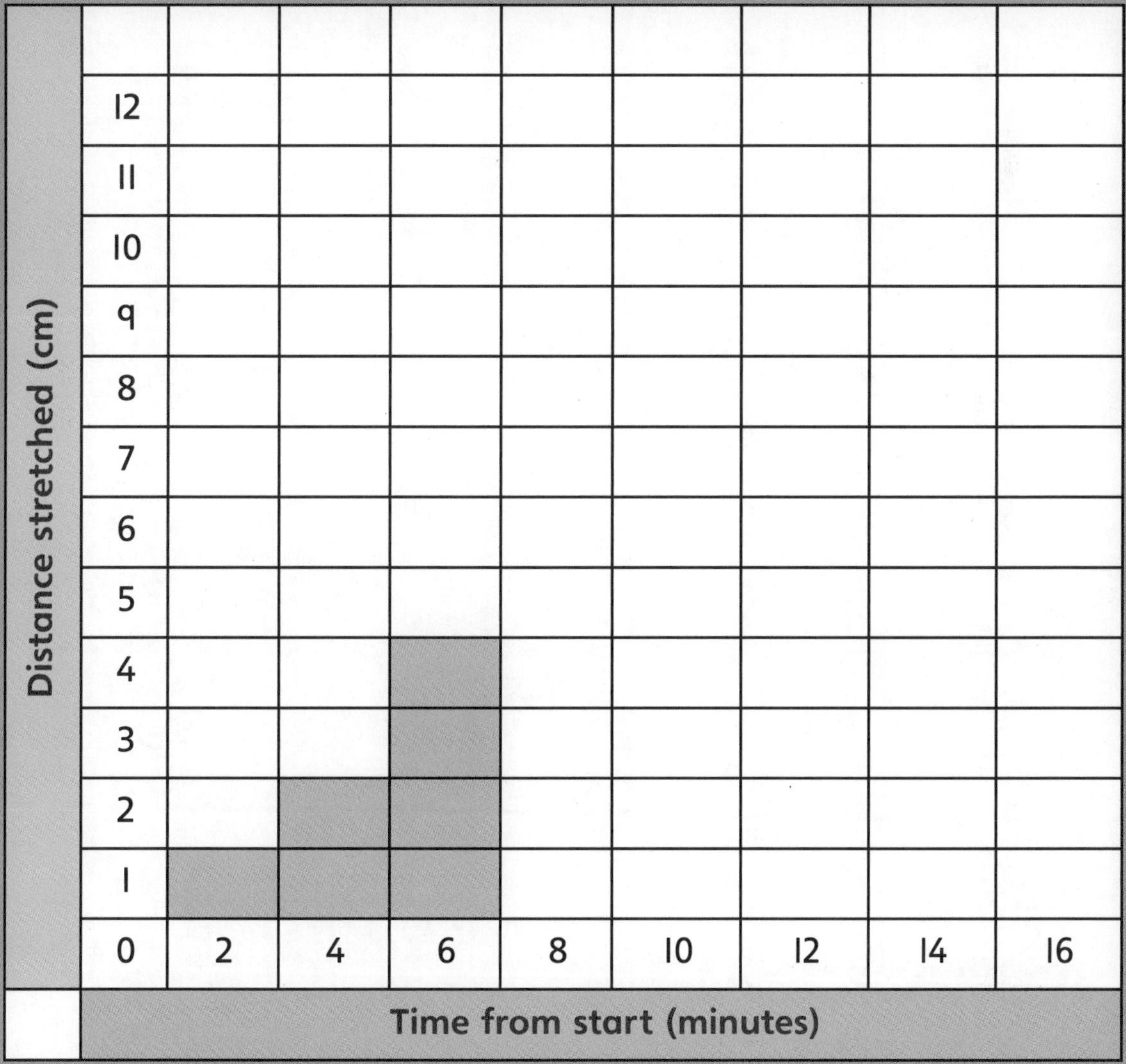

Using Rubber

Rubber is used today to make thousands of things that we use every day.

How many things can you think of that are made from rubber?
Use words and pictures to record your ideas.

Bendy

This bridge is made of strong metal that does not bend. Imagine the bridge is made of a bendy material such as rubber.

1. What would happen when cars and lorries drive over it?

2. Where else is it a bad idea to use bendy materials? Draw a picture of something that you think should not be bendy.

3. What would happen if this object was bendy?

Science Skills

Bendy Ruler - Predict it!

1 Which ruler do you think is the most bendy?

___

2 Put the rulers in order from most bendy to least bendy. Draw them here.

Most bendy	Least bendy

Work with a partner to test the rulers. Carefully bend the rulers. If you bend them too much they might snap! Talk with your partner about your results.

3 What did you discover?
   Did you get the rulers in the right order?
   What surprised you when you tested the rulers?

___
___

Science Skills

Exploring Materials – Record it!

Twisting, weaving and knitting are often used to make fabric. Use a hand lens or microscope to look closely at your clothes. Do they all look the same?

Record what you see in the spaces below and write the name of each material you observe.

Material 1:

Material 2:

Material 3:

Paper Weaving

You will need two different coloured sheets of paper. Follow the steps shown here to weave a paper mat of your own!

Take a piece of paper and fold it in half.

Mark 2 cm spaces along the folded edge.

Cut straight lines from the folded edge almost to the end opposite edge. Do not cut all the way across.

Cut the other piece of paper into strips of approx. 2 cm wide.

Open out the first piece of paper. Take strips of the second paper and weave them through the slots on the first piece of paper.

Use more strips and keep weaving until your mat is complete.

Science Skills

Amazing Balloons - Investigate it!

You can change the shape of a balloon in lots of different ways. Did you know that if you sit carefully on a blown up balloon it will not pop? Balloons are very squashy!

Let's carry out a test to find out which other materials will squash.

1   Write or draw the objects you test in the spaces below, sorting them by whether they squashed or not.

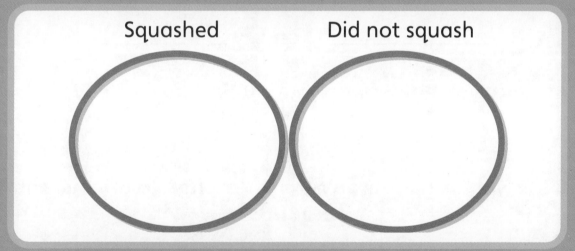

2   Which materials squashed the most?

3   Which materials changed the least?

Making Modelling Dough

Use the words at the bottom of the page to help complete the sentences.

1. Mix the _____ and the _____ together.

2. Add _____. If you add too little the dough will be _____ and crumbly. If you add too much the dough will be too _____.

3. Mix the ingredients until the dough becomes _____ and _____.

4. Once you have made your model you can leave it to _____.

5. It will take a few days for the dough to dry out and _____.

6. Once dry you can use _____ to decorate your models.

dry   flour   harden   paint   salt
soft   squashy   sticky   water

Odd One out

Look at the objects below. For each question draw a circle around the odd one out.

1

Explain your answer.

_____
_____

2

Explain your answer.

_____
_____

3

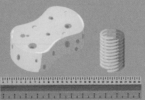

Explain your answer.

_____
_____

What I Have Learned

Think about what you have learned in this unit.

1 What was the most interesting part and why?

2 What else would you have liked to find out about?

Glossary

bend: to curve the shape of a straight object. Something is bendy when it can be easily bent

bounce: describes how an object can spring back from a different surface, such as a floor or wall

cotton: a natural material that grows on a plant and is used to make clothes

harden: changes from soft to hard

latex: white stretchy liquid or solid

metal: materials that are usually hard and strong

mould: used to help shape materials

plant fibre: material that is made from threads or can be used to make rope or yarn

rope: long twisted strands of string-like material

rubber: a stretchy, bouncy material

soft: not hard or stiff

squash: pressing or pushing material so that it changes shape or gets smaller

stretch: pull to make something longer

twist: winding around of an object, and used to join two or more long thin pieces of material together

yarn: thread made from natural and manufactured fibres

词汇表

弯曲：将直线型的物体卷曲，物体容易被弯曲就说明这个物体是柔韧的

反弹：描述一个物体从其他表面（如地板、墙）回弹的过程

棉花：植物生产出的一种自然材料，能用来制作衣服

变硬：从软的变为硬的

乳胶：白色的、有延展性的液体或固体

金属：通常较为坚硬的材料

模具：帮助改变材料形状的物体

植物纤维：由细线状物质形成的材料，能用来制作绳子或纱

绳子：长长的，缠绕在一起的线状材料

橡胶：有延展性和弹性的材料

柔软的：不坚硬的

挤压：将材料按压或推压，让其变形或变小

拉伸：将某物拉得更长

扭转：将物体扭转弯曲，可用来把两个或两个以上的细长材料拧到一起

纱线：由植物纤维和人造纤维制成的细线

培生科学虫双语百科

奇妙物理

Magnets and Forces

磁铁与力

英国培生教育出版集团 著·绘

徐 昂 译

电子工业出版社
Publishing House of Electronics Industry
北京·BEIJING

Original edition, entitled SCIENCE BUG and the title Magnets and Forces Topic Book, by Tanya Shields published by Pearson Education Limited © Pearson Education Limited 2018
ISBN: 9780435196660

All rights reserved. No part of this book may be reproduced or transmitted in any form or by any means, electronic or mechanical, including photocopying, recording or by any information storage retrieval system, without permission from Pearson Education Limited.

This adaptation of SCIENCE BUG is published by arrangement with Pearson Education Limited.

Chinese Simplified Characters and English language (Bi-lingual form) edition published by PUBLISHING HOUSE OF ELECTRONICS INDUSTRY, Copyright © 2023.

For sale and distribution in the mainland of China exclusively (except Hong Kong SAR, Macau SAR and Taiwan).

本书中英双语版由Pearson Education（培生教育出版集团）授权电子工业出版社在中华人民共和国境内（不包括香港、澳门特别行政区及台湾地区）独家出版发行。未经出版者书面许可，不得以任何方式抄袭、复制或节录本书中的任何部分。

本套书封底贴有Pearson Education（培生教育出版集团）激光防伪标签，无标签者不得销售。

版权贸易合同登记号　　图字：01-2022-2381

图书在版编目（CIP）数据

培生科学虫双语百科. 奇妙物理. 磁铁与力：英汉对照 / 英国培生教育出版集团著、绘；徐昂译. --北京：电子工业出版社，2024.1
ISBN 978-7-121-45132-4

Ⅰ.①培… Ⅱ.①英… ②徐… Ⅲ.①科学知识–少儿读物–英、汉 ②磁铁–少儿读物–英、汉 ③力学–少儿读物–英、汉 Ⅳ.①Z228.1 ②O441.3-49 ③O3-49

中国国家版本馆CIP数据核字（2023）第035076号

责任编辑：李黎明　文字编辑：王佳宇
印　　刷：河北迅捷佳彩印刷有限公司
装　　订：河北迅捷佳彩印刷有限公司
出版发行：电子工业出版社
　　　　　北京市海淀区万寿路173信箱　邮编：100036
开　　本：787×1092　1/16　印张：35　字数：840千字
版　　次：2024年1月第1版
印　　次：2024年2月第2次印刷
定　　价：199.00元（全9册）

凡所购买电子工业出版社图书有缺损问题，请向购买书店调换。若书店售缺，请与本社发行部联系，联系及邮购电话：（010）88254888，88258888。
质量投诉请发邮件至zlts@phei.com.cn，盗版侵权举报请发邮件至dbqq@phei.com.cn。
本书咨询联系方式：010-88254417，lilm@phei.com.cn。

使用说明

欢迎来到少年智双语馆！《培生科学虫双语百科》是一套知识全面、妙趣横生的儿童科普丛书，由英国培生教育出版集团组织英国中小学科学教师和教研专家团队编写，根据英国国家课程标准精心设计，可准确对标国内义务教育科学课程标准（2022年版）。丛书涉及物理、化学、生物、地理等学科，主要面向小学1~6年级，能够点燃孩子对科学知识和大千世界的好奇心，激发孩子丰富的想象力。

本书主要内容是小学阶段孩子需要掌握的物理知识，含9个分册，每个分册围绕一个主题进行讲解和练习。每个分册分为三章。第一章是"科学虫趣味课堂"，这一章将为孩子介绍科学知识，培养科学技能，不仅包含单词表、问题和反思模块，还收录了多种有趣、易操作的科学实验和动手活动，有利于培养孩子的科学思维。第二章是"科学虫大闯关"，这一章是根据第一章的知识点设置的学习任务和拓展练习，能够帮助孩子及时巩固知识点，准确评估自己对知识的掌握程度。第三章是"科学词汇加油站"，这一章将全书涉及的重点科学词汇进行了梳理和总结，方便孩子理解和记忆科学词汇。

2024年，《培生科学虫双语百科》系列双语版由我社首次引进出版。为了帮助青少年读者进行高效的独立阅读，并方便家长进行阅读指导或亲子共读，我们为本书设置了以下内容。

（1）每个分册第一部分的英语原文（奇数页）后均配有对应的译文（偶数页），跨页部分除外。读者既可以进行汉英对照阅读，也可以进行单语种独立阅读。问题前面的 📖 符号表示该问题可在第二部分预留的位置作答。

（2）每个分册第二部分的电子版译文可在目录页扫码获取。

（3）本书还配有英音朗读音频和科学活动双语视频，也可在目录页扫码获取。

最后，祝愿每位读者都能够享受双语阅读，在汲取科学知识的同时，看见更大的世界，成为更好的自己！

电子工业出版社青少年教育分社
2024年1月

Contents 目录

Part 1　科学虫趣味课堂　　　　　　　　　　　　　　/ 1
Part 2　科学虫大闯关　　　　　　　　　　　　　　　/ 31
Part 3　科学词汇加油站　　　　　　　　　　　　　　/ 47

Part2译文　　配套音视频

Magnets and Forces

Which forces are being used in this picture?

📖 1a Look at the diagram and then make a list of things you can see in the diagram that move.

📖 1b What makes each of them move?

磁铁与力

图中有几种力？

📖 **1a** 观察图片，请列出你看到的图片中移动的物体。

📖 **1b** 是什么让这些物体移动的？

Moving Toys

We use **forces** every day to make things move. A force is any **push** or **pull** that makes something move.

Word Box
force
pull
push

1 Look at the toys in the shop window. How would you make the different toys move?

2a Make a list of any toys you have played with that need a push force to make them move.

2b Make a list of toys that need a pull force to make them move.

3 Make a list of other activities that use forces to make things move.

移动的玩具

我们每天都会使用**力**来让物体移动。力指的是能让物体移动的**推力**或**拉力**。

单词表
force 力
pull 拉力
push 推力

1 观察图片中商店橱窗里的玩具。怎样才能让不同的玩具移动?

📖 **2a** 请列出你玩过的需要推力才能移动的玩具。

📖 **2b** 请列出需要拉力才能移动的玩具。

📖 **3** 还有哪些活动也要运用力的作用来移动物体,请列出来。

Pushes and Pulls

Some things are designed to be both pushed and pulled. We use a pull force to open drawers. We use a push force to close drawers.

We pull a drawer to open it.

We push a drawer to close it.

1 Make a list of things that use both push and pull forces.

Forces are used to make the swing move.

2 What kind of force is needed to make the swing move?

3 How would you slow the swing down?

推力和拉力

人们设计出一些既能被推动又能被拉动的物体。我们用拉力来打开抽屉，用推力来关上抽屉。

我们通过"拉"的动作来打开抽屉。

我们通过"推"的动作来关上抽屉。

1 请列出需要同时用到推力和拉力的物体。

力能让秋千荡起来。

2 让秋千荡起来需要哪种力？

3 如何让秋千慢下来？

Science Skills

Measure it!

Word Box
measure

Look at the diagram to find out how the car catapult works.

Nadia, Khalid and Carlos decided to pull back the elastic band to different distances to **measure** how far the car would travel when they let go of the elastic band.

Look at Nadia's, Khalid's and Carlos's ideas.

Let's do a small pull and a big pull on the elastic band.

Let's use 1 cm cubes to see how far back we pull the elastic band.

I think we should use a 30 cm ruler to measure the pull back.

科学技能

测量一下吧！

单词表
measure 测量

观察图片，探究一下汽车弹弓是如何工作的。

纳迪娅、哈立德和卡洛斯决定将皮筋向后拉伸不同的距离，测量放开皮筋后汽车移动的距离。

一起来看看纳迪娅、哈立德和卡洛斯的想法吧。

让我们分别用较小的力和较大的力来拉伸皮筋吧。

让我们用一些棱长为1厘米的立方体来测量皮筋被我们拉到多远吧。

我认为我们应该用长度为30厘米的直尺来测量皮筋被拉伸的距离。

The children also wanted to know how far the car moved forwards.

I think we should make a mark on the floor to show how far each car travels.

I think we should use a metre ruler to measure how far the car travels.

I think we should use a tape measure because it might go further than a metre.

📖 Test it yourself

Make your own car catapult and measure how far the car travels when you change the distance you pull back the elastic band.

小朋友们还想知道玩具汽车向前移动了多远。

我认为我们应该在地板上做出标记,这样可以记录玩具汽车每次移动的距离。

我认为我们应该用米尺来测量玩具汽车移动的距离。

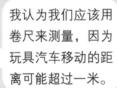

我认为我们应该用卷尺来测量,因为玩具汽车移动的距离可能超过一米。

自己试一试

自己制作一个汽车弹弓,改变皮筋被拉伸的长度,然后在皮筋不同的拉伸长度下,测量汽车移动的距离。

Boules

Boules is a game, where the aim of the game is to roll balls towards a smaller target ball. The person who gets their ball the closest to the target ball is the winner.

Word Box
gravel
surface

The game is usually played on a hard **gravel surface**.

1 What do you think would happen if you played boules on a smooth surface?

2 What do you think would happen if you played boules on a grassy surface?

3 Try rolling balls on different surfaces. What do you notice?

The larger balls are usually quite heavy.

People have been playing boules for thousands of years, since the time of the ancient Greeks.

地掷球

地掷球是一种游戏，人们需要将球掷向更小的目标球。离目标球最近的人是获胜者。这个游戏通常在坚硬的**碎石表面**上进行。

单词表
gravel 碎石
surface 表面

1 如果在光滑的表面上进行地掷球游戏，你认为会发生什么？
2 如果在长满草的表面上进行地掷球游戏，你认为会发生什么？
3 试着在不同材质的表面上抛掷球，你发现了什么？

目标球

球越大通常就越沉。

从几千年前的古希腊时期起，人们就开始玩地掷球了。

Design a Table Top Game

Think about how people play boules. They use forces to move balls towards a target object. Design a game that can be played on a table and uses similar rules to boules.

Design your own game

You will need objects to use instead of boules, e.g. coins, marbles or small stones. You will also need thick card to make the playing board and different playing surfaces, e.g. sandpaper, card, fabric or tissue.

1. Design a game so it fits onto a single table top. The target object should be at least 60 cm from the start line.
2. Carry out a test to find out which materials are the best for making a table top game.

1. Make a list of the objects needed to play your game.
2. Write a set of instructions and list some helpful hints for playing your game.

设计一个桌游

思考一下人们是如何玩地掷球的。他们运用力的作用让球向目标球的方向移动。请按照地掷球类似的规则,设计一个可以在桌面上玩的游戏。

设计自己的游戏

你将需要代替地掷球的物体,例如,硬币、弹珠或者小石头。你还需要厚纸板以及材质不同的材料,例如,砂纸、卡片、布料或纸巾,将它们作为游戏桌面。

1. 设计一个在单个桌面完成的游戏。目标物至少应该距离起点线60厘米。
2. 进行实验,确定最适合用于桌游的材料。

1. 请列出在玩你设计的游戏时需要的物体。
2. 请写出游戏说明,并列出一些对玩游戏有帮助的提示。

Magnets

Magnets are special objects that use a force called **magnetism** to **attract** other magnets and some **metals**.

Word Box
attract
magnetism
metal
pole
repel

This is a bar magnet.

Magnets have two magnetic **poles**. We call these poles the north pole and the south pole.

1 Make a list of where you see magnets being used. For example, fridge doors have magnets to keep them closed.

2 Do all magnets look the same? Explain your answer.

3 Draw the different types of magnets you have seen.

All magnets have a north pole and a south pole.

磁铁

磁铁是用**磁力吸引**其他磁体和某些**金属**的特殊物体。

单词表
attract 吸引
magnetism 磁力
metal 金属
pole 极
repel 排斥

这是一块条形磁铁。

磁铁有两个磁**极**。我们把这两个磁极分别叫作北极和南极。

1 请列出你见过的应用了磁铁的场景。例如,冰箱门上有磁铁才可以让门合上。

2 所有磁铁的形状都一样吗?请解释你的回答。

3 请画出你见过的不同种类的磁铁。

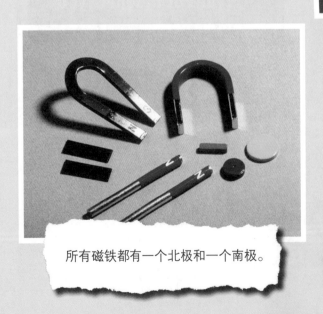

所有磁铁都有一个北极和一个南极。

Explore what happens when you move two magnets close to each other.

investigating magnets

You will need: two bar magnets.

1. What happens when you move the north pole on a magnet towards the south pole on another magnet?
2. What happens when you move the north pole on a magnet towards the north pole on another magnet?
3. What happens when you move the south pole on a magnet towards the south pole on another magnet?

When two magnets pull towards each other we say they attract each other.

When two magnets push away from each other we say they **repel** each other.

Opposite poles attract each other.

The same poles repel each other.

当你让两块磁铁靠近时，探索一下会发生什么。

探究磁铁

你将需要：两块条形磁铁。

1. 当你让一块磁铁的北极靠近另一块磁铁的南极时，会发生什么？
2. 当你让一块磁铁的北极靠近另一块磁铁的北极时，会发生什么？
3. 当你让一块磁铁的南极靠近另一块磁铁的南极时，会发生什么？

如果两块磁铁之间产生拉力，我们就说这两块磁铁相互吸引。
如果两块磁铁之间产生推力，我们就说这两块磁铁相互**排斥**。

异极相互吸引。

同极相互排斥。

Magnetic Materials

Magnets can attract other **materials**. Materials that are attracted to magnets are **magnetic**. Look at the materials below.

Word Box
iron
magnetic
material
steel

Testing which materials are magnetic

You will need: a magnet, the objects in the pictures...

1 Use your magnet to do a test to find materials that are magnetic.

2 Record which materials are magnetic and which materials are not.

3 What do you notice about the materials in the magnetic list?

磁性材料

磁铁能吸引其他**材料**。
能受到磁铁吸引的材料是**有磁性的**。
观察下面的材料。

单词表

iron 铁
magnetic 有磁性的
material 材料；物质
steel 钢

测试一下哪些材料是有磁性的

你将需要：一块磁铁、图片中的物体……

1 请你用磁铁做一项测试，找出哪些材料是有磁性的。

2 记录哪些材料是有磁性的，哪些材料是没有磁性的。

3 在记录的有磁性的材料中，你注意到了什么？

Iron is a metal that is used to make **steel**. All materials that contain iron are attracted to magnets. We call them magnetic materials.

These objects are all made from metals that contain iron. They are all magnetic.

Search for magnetic materials

1. Use a magnet to find metal objects that are magnetic. Do not place magnets near electrical objects or computer screens as they can damage them.
2. Make a list of all the objects you find.

Some metals such as gold, aluminium and copper are not attracted to magnets. This means they are not magnetic.

This gold ring, aluminium foil and copper pipe are all made from non-magnetic metals.

铁是一种被用于制成**钢**的金属。所有含有铁的材料都能被磁铁吸引。我们把这些材料叫作磁性材料。

这些物体都是由含铁的金属制成的。它们都是有磁性的。

寻找磁性材料

1. 用一块磁铁来寻找有磁性的金属物体。不要将磁铁靠近电力设备或电脑屏幕，因为磁铁会损坏它们。
2. 请列出你找到的所有带磁性的物体。

一些金属不受磁铁吸引，例如，金、铝、铜。这意味着它们没有磁性。

金戒指、铝箔纸和铜管都是由没有磁性的金属制成的。

Magnetic Force

A **magnetic force** is different from the forces we have looked at so far. A magnetic force does not need to touch objects for a push or a pull to happen. A magnetic force can attract (pull) or repel (push) objects without touching them.

Word Box
magnetic force

1. Why is the paperclip held in the air?
2. What do you think will happen if you place a piece of card between the paperclip and the magnet?

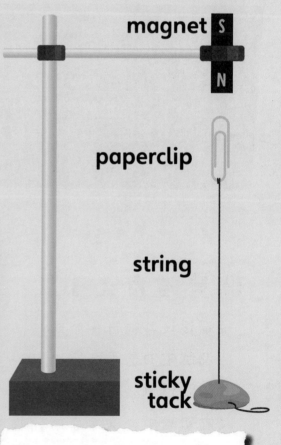

The magnet is not touching the paperclip.

Testing magnetic force

1. Plan and carry out your own test to find out if a magnetic force can pass through different materials.
2. What equipment will you need for your test?
3. What did you find out?

磁力

磁力不同于我们前面提到的其他力。磁力不需要物体间相互接触,就能产生推力或拉力。在不接触物体的情况下,磁力就可以吸引(拉)或排斥(推)物体。

单词表
magnetic force 磁力

📖 1 为什么图中的回形针悬在空中?

📖 2 如果将一张纸放在回形针和磁铁之间,你认为会发生什么?

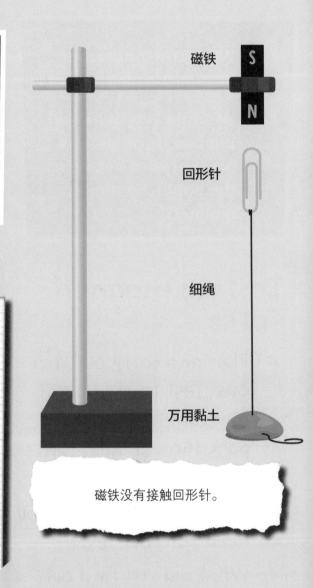

磁铁
回形针
细绳
万用黏土

磁铁没有接触回形针。

测量磁力大小

1 请设计并进行一个实验,测试磁力是否能穿过不同的材料。

2 在实验中,你需要用到哪些设备?

📖 3 在实验中,你发现了什么?

Strongest Magnet

Magnets come in lots of different shapes and sizes. The shape and size of a magnet can change the strength of the magnetic force.

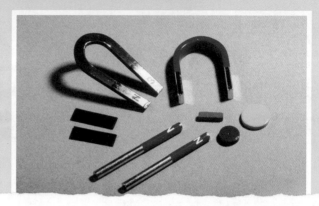

There are lots of different types of magnet.

Testing the strength of magnets

You will need: some magnets, magnetic paperclips...

1. Collect some different magnets. Which one do you think is the strongest?
2. Plan and carry out a test to find out which magnet is the strongest.
3. Move the magnet along the ruler until the paperclip moves towards the magnet.
4. Use the ruler to measure the distance the magnetic force starts to attract the paperclip.

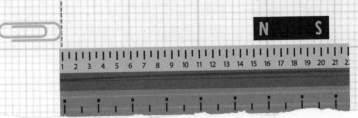

The stronger the magnetic force the bigger the distance will be between the paperclip and the magnet.

磁力最大的磁铁

磁铁有不同的形状和尺寸。一块磁铁的形状和尺寸可以影响它产生的磁力的大小。

磁铁有很多不同的种类。

测量磁铁的磁力大小

你将需要：几块磁铁、有磁性的回形针……

1. 收集几块不同的磁铁。你认为哪一块磁铁的磁力最大？
2. 设计并进行一个实验，找出磁力最大的磁铁。
3. 将磁铁沿着直尺缓慢移动，直到回形针被磁铁吸引。
4. 使用直尺测量回形针开始受到磁力作用的距离。

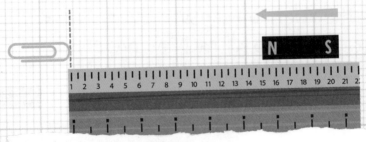

磁力越大，回形针和磁铁之间的距离就越远。

A Magnetic Discovery

This ancient Greek legend tells the story of how a man called Magnes discovered magnets.

Word Box
compass

One day, Magnes was moving his sheep. His boots began to feel heavier and heavier. He stopped to see what was happening. Magnes found rocks that seemed to 'stick' to the iron nails in his boots.

Magnes told people about his discovery and they began collecting the rocks for themselves. No one could think of a use for them, but they liked to play with them.

People believed that the rocks were magical.

发现磁铁

古希腊神话中讲述了一位名叫马格内斯（Magnes）的人发现磁铁的故事。

单词表
compass 指南针

一天，马格内斯在放羊。他感觉自己的靴子变得越来越重。于是，他停下来观察靴子到底怎么了。马格内斯发现，石头似乎"粘"在了靴子的铁钉上。

马格内斯将这个发现告诉了其他人，人们也开始收集这种石头。没人能想出磁石的具体作用，但是，他们喜欢玩磁石。

人们相信这种石头是有魔力的。

Hundreds of years ago people in the UK used to call magnetic rocks 'lodestones', from the Middle English 'lode' meaning 'guide'.
The Earth's North Pole is magnetic and attracts other magnets. Travellers and sailors used to use lodestones suspended on a piece of rope, to identify which way was north. The lodestone would turn to point north. This helped them to find their way when they travelled to faraway places.
Many years later people used magnets to make **compasses**.

Today we use magnets in all sorts of things. Recycling companies use magnets to sort magnetic and non-magnetic waste materials. A large magnet is passed over the different objects and all the magnetic materials are attracted to the magnet.

People use compasses to help them find out which way is north.

Magnetic materials sticking to a large magnet in a recycling centre.

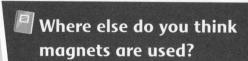

Where else do you think magnets are used?

几百年前，英国人将磁石称作"天然磁石（lodestone）"，"lode"一词在中世纪英文中意为"指引"。

地球的北极是有磁性的，会吸引其他磁体。旅行者和水手们过去常常将天然磁石悬挂在一根绳索上，用来辨别哪一面是北。磁石会指向北方，这帮助人们在长途跋涉时找到正确的方向。许多年之后，人们利用磁铁制作了**指南针**。

今天，我们在各种地方都能用到磁铁。垃圾回收站利用磁铁对有磁性和无磁性的垃圾进行分类。一块大磁铁从垃圾上空缓慢移动，所有有磁性的物质都被吸引到了大磁铁上。

人们利用指南针来帮助他们找到北方。

垃圾回收站里，磁性物质被吸引到一大块磁铁上。

📖 你还能想到人们利用磁铁的其他场合吗？

Magnets and Forces

1a Think about the things that move in a playground, then draw pictures of things you thought of.

1b Add labels to your picture to show what makes each of them move.

Moving Toys

1. Think about the toys you have played with. List the toys that need a push to make them move and the toys that need a pull to make them move.

Toys that need a push to make them move	Toys that need a pull to make them move

2. Make a list of other activities that use forces to make things move.

Pushes and Pulls

1. Draw a picture of things that use both push and pull forces.

2. What kind of force is needed to make the swing move?

3. How would you slow the swing down?

Science Skills

Measure it!

Look at the children's ideas about how to pull back the elastic band in their car catapult investigation.

The children also wanted to know how far the car moved forwards.

> Let's do a small pull and a big pull on the elastic band.

> Let's use 1 cm cubes to see how far back we pull the elastic band.

> I think we should use a 30 cm ruler to measure the pull back.

1 Which idea do you think is the best and why?

_____
_____
_____

> I think we should use a tape measure because it might go further than a metre.

> I think we should make a mark on the floor to show how far each car travels.

> I think we should use a metre ruler to measure how far the car travels.

2 Which idea do you think is the best and why?

_____
_____
_____

3 Plan and carry out your own test and record the results in the table below.

The distance you pull the elastic band back from the start line (cm)	The distance the car travels (cm)

4 What patterns do you notice in your results?

5 How could you improve your investigation?

Boules

Boules is a game usually played on hard gravel surfaces.

1. What do you think would happen if you played boules on a smooth surface?

   _____

2. What do you think would happen if you played boules on a grassy surface?

   _____

3. Try rolling balls on different surfaces. What do you notice?

   _____
   _____

4. Make a list of other sports that need a smooth surface to play on.

   _____   _____

   _____   _____

   _____   _____

Design a Table Top Game

1 Design a game that can be played on a table and uses similar rules to boules. Draw a plan of your game in the space below.

2 Make a list of the objects needed to play your game.

3 Write a set of instructions and list some helpful hints for playing your game.

Magnets

1. Make a list of where you see magnets being used. You can use pictures and words to answer this question.

2. Do all magnets look the same? Explain your answer.

3. Draw the different types of magnets you have seen.

4 Draw a diagram to show what happens when you move the north pole on a magnet towards the south pole on another magnet.

5 Draw a diagram to show what happens when you move the north pole on a magnet towards the north pole on another magnet.

6 Draw a diagram to show what happens when you move the south pole on a magnet towards the south pole on another magnet.

Magnetic Materials

1. Look at the materials above. Record which materials you think might be magnetic and those you think are not magnetic.

Magnetic	Not magnetic

Use a magnet to find out which materials are magnetic and which materials are not magnetic.

2. Look at your predictions in the table above. Draw a circle around the materials that should be grouped in a different column.

3 What do you notice about the materials in the magnetic list?

4 Use a magnet to find metal objects that are magnetic. Make a list of all the objects you find. You can use pictures and words to answer this question.

5 Make a list of metal objects that are not magnetic. You can use pictures and words to answer this question.

Magnetic Force

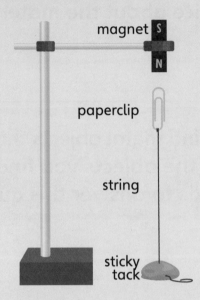

1   Why is the paperclip held in the air?

2   If you place a piece of card between the paperclip and the magnet what do you think would happen?

3   Set up your own investigation to find out if a magnetic force can pass through different materials.

   What did you find out? Record your observations in the table.

Material	Observation

Strongest Magnet

1. Collect some different magnets. Draw a picture of the different magnets in the space below.

2. Which magnet do you think will be the strongest?

3. Plan and carry out a test to find out which magnet is the strongest. How did you set up your test to find the strongest magnet?

4. Which magnet is the strongest?

5. Which magnet is the weakest?

A Magnetic Discovery

An ancient Greek legend tells the story of how a man called Magnes discovered magnets.

1 Create a comic strip to retell the story of how magnets were first discovered.

1.	2.
3.	4.

Today we use magnets in lots of different ways.

2 Research the different ways magnets are used and record your answers in the space below. You can use pictures and words to answer this question.

What I Know About Magnets and Forces

Use the traffic lights to record what you have learned and how well you completed the tasks.

	I enjoyed learning about…	My best piece of work was…
	It was OK learning about…	I need to make some small improvements to my work on…
	I did not enjoy leaning about…	The piece of work that needs the most improvement is…

Glossary

attract: a pull force between two magnets, so they move towards each other

compass: a device that identifies which direction is north

force: pushes and pulls that can be used to make things move

gravel: small rough stones often used to cover paths

iron: a strong, hard grey metal that is magnetic

magnetic: materials that are attracted to a magnet

magnetic force: the force that magnets have. It can repel or attract magnetic materials

magnetism: the force that magnets have to repel or attract magnetic materials

measure: a scientific way to find the length or mass of something

metal: a type of material that is usually strong, shiny and sometimes magnetic

material: what objects are made of

pole: the two ends of a magnet (north and south)

pull: a force that makes things move towards something

push: a force that makes things move away from something

repel: a push force between two magnets, so they move away from each other

steel: a strong metal made up of different materials including iron

surface: the outer part of an object

词汇表

吸引：两块磁铁之间产生的拉力，能让两块磁铁相向移动

指南针：可以辨别北方的装置

力：能让物体移动的推力和拉力

碎石：小块而不平滑的石头，常用来铺路

铁：有磁性的硬质灰色金属

有磁性的：能够被磁铁吸引的物质材料

磁力：磁铁特有的作用力，可以排斥或吸引磁性物质

磁性：磁铁排斥或吸引磁性物质的特性

测量：一种科学研究方法，可以测量物体的长度或质量

金属：通常是坚硬的且有光泽的，有时还具有磁性的物质

材料：物体的具体组成

极：一块磁铁的两端（北极和南极）

拉力：让物体移近的力

推力：让物体移远的力

排斥：两块磁铁之间的推力，能让磁铁向相反的方向移动

钢：一种含有铁和其他物质的坚硬的合金

表面：物体的外部

培生科学虫双语百科
奇妙物理

Light and Shadows

光与影

英国培生教育出版集团 著·绘
徐 昂 译

电子工业出版社
Publishing House of Electronics Industry
北京·BEIJING

Original edition, entitled SCIENCE BUG and the title Light and Shadows Topic Book, by Deborah Herridge published by Pearson Education Limited © Pearson Education Limited 2018
ISBN: 9780435196493

All rights reserved. No part of this book may be reproduced or transmitted in any form or by any means, electronic or mechanical, including photocopying, recording or by any information storage retrieval system, without permission from Pearson Education Limited.

This adaptation of SCIENCE BUG is published by arrangement with Pearson Education Limited. Chinese Simplified Characters and English language (Bi-lingual form) edition published by PUBLISHING HOUSE OF ELECTRONICS INDUSTRY, Copyright © 2023.

For sale and distribution in the mainland of China exclusively (except Hong Kong SAR, Macau SAR and Taiwan).

本书中英双语版由Pearson Education（培生教育出版集团）授权电子工业出版社在中华人民共和国境内（不包括香港、澳门特别行政区及台湾地区）独家出版发行。未经出版者书面许可，不得以任何方式抄袭、复制或节录本书中的任何部分。

本套书封底贴有Pearson Education（培生教育出版集团）激光防伪标签，无标签者不得销售。

版权贸易合同登记号　　图字：01-2022-2381

图书在版编目（CIP）数据

培生科学虫双语百科. 奇妙物理. 光与影：英汉对照 / 英国培生教育出版集团著、绘；徐昂译. --北京：电子工业出版社，2024.1
ISBN 978-7-121-45132-4

Ⅰ.①培… Ⅱ.①英… ②徐… Ⅲ.①科学知识－少儿读物－英、汉 ②光学－少儿读物－英、汉 Ⅳ.①Z228.1 ②O43-49

中国国家版本馆CIP数据核字（2023）第035075号

责任编辑：李黎明　文字编辑：王佳宇
印　　刷：河北迅捷佳彩印刷有限公司
装　　订：河北迅捷佳彩印刷有限公司
出版发行：电子工业出版社
　　　　　北京市海淀区万寿路173信箱　邮编：100036
开　　本：787×1092　1/16　印张：35　字数：840千字
版　　次：2024年1月第1版
印　　次：2024年2月第2次印刷
定　　价：199.00元（全9册）

凡所购买电子工业出版社图书有缺损问题，请向购买书店调换。若书店售缺，请与本社发行部联系，联系及邮购电话：（010）88254888，88258888。
质量投诉请发邮件至zlts@phei.com.cn，盗版侵权举报请发邮件至dbqq@phei.com.cn。
本书咨询联系方式：010-88254417，lilm@phei.com.cn。

使用说明

欢迎来到少年智双语馆！《培生科学虫双语百科》是一套知识全面、妙趣横生的儿童科普丛书，由英国培生教育出版集团组织英国中小学科学教师和教研专家团队编写，根据英国国家课程标准精心设计，可准确对标国内义务教育科学课程标准（2022年版）。丛书涉及物理、化学、生物、地理等学科，主要面向小学1~6年级，能够点燃孩子对科学知识和大千世界的好奇心，激发孩子丰富的想象力。

本书主要内容是小学阶段孩子需要掌握的物理知识，含9个分册，每个分册围绕一个主题进行讲解和练习。每个分册分为三章。第一章是"科学虫趣味课堂"，这一章将为孩子介绍科学知识，培养科学技能，不仅包含单词表、问题和反思模块，还收录了多种有趣、易操作的科学实验和动手活动，有利于培养孩子的科学思维。第二章是"科学虫大闯关"，这一章是根据第一章的知识点设置的学习任务和拓展练习，能够帮助孩子及时巩固知识点，准确评估自己对知识的掌握程度。第三章是"科学词汇加油站"，这一章将全书涉及的重点科学词汇进行了梳理和总结，方便孩子理解和记忆科学词汇。

2024年，《培生科学虫双语百科》系列双语版由我社首次引进出版。为了帮助青少年读者进行高效的独立阅读，并方便家长进行阅读指导或亲子共读，我们为本书设置了以下内容。

（1）每个分册第一部分的英语原文（奇数页）后均配有对应的译文（偶数页），跨页部分除外。读者既可以进行汉英对照阅读，也可以进行单语种独立阅读。问题前面的 📖 符号表示该问题可在第二部分预留的位置作答。

（2）每个分册第二部分的电子版译文可在目录页扫码获取。

（3）本书还配有英音朗读音频和科学活动双语视频，也可在目录页扫码获取。

最后，祝愿每位读者都能够享受双语阅读，在汲取科学知识的同时，看见更大的世界，成为更好的自己！

电子工业出版社青少年教育分社
2024年1月

Contents 目录

Part 1　科学虫趣味课堂　　　　　　　　　　　　　　/ 1

Part 2　科学虫大闯关　　　　　　　　　　　　　　　/ 33

Part 3　科学词汇加油站　　　　　　　　　　　　　　/ 49

Part2译文

配套音视频

Where Does Light Come from?

Word Box
absence
dark
light
source
Sun

Is it **light** or **dark** where you are now?
What lights can you see?
Things that make light are called **sources** of light.

> **You will need:**
> old magazines, leaflets, newspapers, catalogues, scissors, glue, some plain paper...
> Draw or cut some pictures of sources of light. How many can you find? Make them into a poster.

The **Sun is** our main source of light.

We are surrounded by sources of light. Without light we cannot see. Where there is no light we say it is dark. Darkness is the **absence** of light.

We can hardly ever see real darkness in towns. There are so many lights at night from street lights, car headlights, shop windows and houses that it is never really dark.

This city is so light at night we cannot see the stars!

光从哪里来？

你现在所在的地方是**亮的**还是**暗的**？
你能看见哪些光？
能发光的物质叫作**光源**。

> **单词表**
> absence 缺失；没有
> dark 暗的；黑暗的
> light 光；亮的
> source 来源；光源
> Sun 太阳

你将需要：
旧杂志、传单、报纸、目录簿、剪刀、胶水、一些白纸……
画出或者剪一些光源的图片，你能发现多少种？把这些图片制作成一张海报。

太阳是我们主要的光源。

我们被光源围绕着。没有光，我们就看不见。我们将没有光的环境称为黑暗，黑暗就是**没有**光。

在城市里，我们几乎不能见到真正的黑暗，因为路灯、车灯、商店橱窗、楼房等发出了各种各样的光，所以城市里没有真正的黑暗。

城市夜晚的灯光太强，所以我们无法看见星星！

Sources of Light

When it is dark we need light so we can see. These explorers are in dark caves underground. They have lights on their helmets and hold torches so they can see what is in front of them. The helmet lights and torches are sources of light.

What would these explorers see if they turned off their lights?

1 Which of these are sources of light?
2 How do you know?

3 Where is the darkest place you have been?

光源

当我们身处黑暗中时,我们需要借助光才能看见。这些勘探者身处地下黑暗的洞穴,他们头上戴着有灯的头盔,这样他们才能看见前面的事物。头盔上的灯和手电筒就是光源。

如果勘探者们关闭灯光,他们会看见什么?

1 下列图片哪些是光源?
2 你是如何知道的?

3 你去过的最暗的地方是哪里?

Brighter or dimmer

Not all light sources are the same. Some light sources are brighter than others. These floodlights are very bright.

Bright floodlights make it seem like daytime.

This nightlight is dim. It does not make much light.

A brighter light would wake us up!

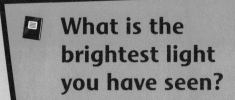

What is the brightest light you have seen?

更亮或更暗

不是所有的光源都是一样的。有些光源比其他光源更亮。这些探照灯就十分明亮。

在明亮的探照灯的照射下，黑夜就如同白昼。

这盏夜间照明灯比较暗，没有发出很多光。

更亮的光就会将我们从睡梦中唤醒啦！

你见过的最亮的光是什么光？

Reflecting Light

Some things do not make their own light. Mirrors and the Moon do not make their own light. They are not light sources. We see them because they **reflect** light. Light from light sources bounces off shiny or pale coloured objects and they appear bright.

The Moon is a **reflector** of light. We see the Moon at night because light from the Sun bounces off it.

Word Box
reflect
reflection
reflector

The Moon is a light reflector.

1 List the sources of light and reflectors in the picture below.

2 What would you see if it were night time in the picture?

Reflective strips on clothing mean we can be seen at night.

反射光

一些物体自身不会发光。平面镜和月亮自身就不会发光，它们不是光源。我们之所以能看见它们，是因为它们会反射光。光源发出的光遇到闪亮的或浅颜色的物体会发生**反射**，导致这些物体变亮了。

月亮就是一个光源的**反光体**，我们能在晚上看到月亮是因为月亮反射了来自太阳的光。

单词表
reflect 反射
reflection 镜像
reflector 反光体

月亮是一个反光体。

1 请列出下图中的光源和反光体。
2 如果图中的场景发生在夜间，你能看见哪些东西？

衣服上的反光条方便我们在晚上能被别人看见。

In the Dark

Some objects are light sources, some reflect light.

1 Which objects do you think make good reflectors?

Make a 'classroom cave'

You could throw a blanket over a table so it is really dark inside. Stick paper or plastic in bags over any gaps where the light gets in.

Collect some objects you think might be sources of light and some that you think might be reflectors.

Take them into the cave one at a time.

Which can you see? What can you do to see the objects that do not make their own light?

We need a light source to see in the dark.

More about reflection

Not all objects reflect light well. Look at the smooth and shiny surface of a mirror. Can you see your **reflection**? Look at the dull, dark surface of a brick wall. Can you see your reflection now?

Which objects reflect light well?

2 How many things can you find that reflect light? What test will you use to check?

3 What do good reflectors all have in common?

在黑暗中

一些物体是光源，一些物体能反射光。

> 1 你认为哪些物体是好的反光体？

制作一个"教室洞穴"

你可以用毯子盖住一张桌子，这样桌子下面会变得很黑。用纸或塑料袋遮住所有漏光的地方，然后收集一些你认为可以作为光源和反光体的物品。

将这些物品都放进洞穴里，你能看见哪个？你需要如何做才能看到本身不会发光的物体？

我们需要光源，这样才能在黑暗中看见。

更多关于反射的信息

不是所有的物体都能很好地反射光。观察表面光滑而闪亮的平面镜，你能看到自己的**镜像**吗？再观察表面阴暗的砖墙，你还能看到自己的镜像吗？

哪些物体能很好地反射光？

> 2 你能找到多少个反光的物体？你是如何检验的？
> 3 好的反光体有什么共同点？

How Does Light Travel?

You know that light can bounce off some materials but how does it travel? Light moves from a light source in a **straight** line. Light can only travel in straight lines.

Light travels in straight lines.

Word Box
straight
light ray

What does the arrow show in the tube diagram?

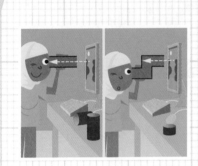

Roll a piece of black card into a tube. Look through it at a light source such as a computer screen. ⚠ Never look at the Sun! Can you see the light?

Now bend the tube. Can you still see the light? Light cannot travel around the bend in the tube so you cannot see any light.

Sometimes you can see straight lines of light shining through trees or clouds. We call these lines **light rays**. Can you see the rays of light from the Sun shining through the gaps in the leaves?

光是如何传播的？

你已经知道光能在一些材料的表面反射了，那么光又是如何传播的呢？光源发出的光沿直线传播，而且光只沿直线传播。

光沿直线传播。

单词表
straight 笔直的
light ray 光线

下图中管子里的箭头表示的是什么？

将一张黑卡片卷起来并放到一根管子内，透过管子观察光源（比如电脑屏幕）。⚠ 不要观察太阳！你能看到光吗？

现在将管子弯曲，你还能看到光吗？光不能经过管子的弯曲处传播，因此你看不到任何光。

有时你能透过树或云看到沿直线传播的照射下来的光。我们把这些光叫作**光线**。你能看到太阳光透过树叶间的缝隙照射下来的光吗？

Bounce a Light

Word Box
beam

You can use the knowledge that light travels in straight lines to play a game.

You will need:

a torch (your light source), 4 friends, each with a small mirror...

Stand the friends in each corner of a darkened room. Shine the torch so the light hits one of the mirrors. Work out how to get the light to bounce from one mirror to the next and back to the beginning. How quickly can you get the light **beam** back to the starting point?

How could you use mirrors to bounce light around a corner?

Make a light ray box

You will need: an old shoe box, black paint, scissors, a torch...

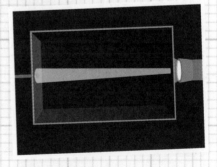

Paint the inside of the shoebox black and allow it to dry.

Cut a small hole exactly in the centre of each short end of the box.

In a dark room, shine a torch through one of the holes. Can you see the path the light takes? Does the light shine through the hole on the opposite end?

反射光

你可以利用光沿直线传播的原理来玩游戏。

单词表
beam 光束

你将需要：

一只手电筒（你的光源）、四个朋友、每个朋友带一面小平面镜……

让你的四个朋友手持小平面镜分别站在黑暗房间的四个角落。打开手电筒，照向其中一面平面镜。试着让光线从一面平面镜反射到另一面平面镜上，依次反射回最初的平面镜上。你需要多长时间让**光束**反射回起点？

你是如何利用平面镜让光在角落反射的？

制作一个光线盒

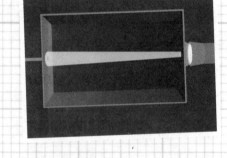

你将需要：一个旧鞋盒、黑漆、一把剪刀、一只手电筒……

将鞋盒的内部刷上黑漆，等待油漆变干。

在鞋盒两个窄面的中心处分别剪出两个小洞。

在黑暗的房间里，使手电筒的光从一个小洞照向另一侧。你能观察到光的传播路径吗？光会穿过小洞到达对面吗？

What Is a Shadow?

Light travels from a light source in straight lines. It cannot travel around corners. When light hits some objects or some materials it is blocked and cannot get through. Where there is no light there is darkness. Objects that block light can make **shadows**. Shadows are a similar shape to the objects that make them.

Word Box
shadow

Find some objects in your classroom. Use a torch to make a shadow of the object. What shape is the shadow?

Shadows outside

The Sun is an important source of light that gives us light in the daytime. When light from the Sun is blocked we get shadows.

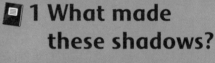 1 What made these shadows?
2 Why are shadows usually black?

Go outside on a sunny day. How many shadows can you find?
Make some shadows with your body. How many shapes can you make?

什么是影子？

光从光源发出且沿直线传播。它不能在转角处变换方向。当光碰到某些物体或某些材料时，光受到阻碍而无法穿过。光到达不了的地方就是黑暗。阻碍光传播的物体可以形成**影子**，影子的形状与形成影子的物体的形状相似。

单词表
shadow 影子

在你的教室里寻找一些物品，用手电筒照射物品形成影子，影子的形状是怎样的？

外面的影子

太阳是重要的光源，能在白天为我们提供光线。当来自太阳的光被阻挡时，我们就能得到影子。

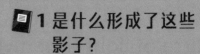

1 是什么形成了这些影子？
2 为什么影子通常是黑色的？

在晴朗的天气里出去走走。你能找到多少种影子？

用你的身体形成一些影子。你能形成多少种不同形状的影子？

Opaque Objects

Some materials block light. When light is blocked you can see a shadow. Objects and materials that do not let light through are **opaque**. Light cannot go through opaque materials or opaque objects. Light around the object is not blocked.

Word Box
cast
opaque
silhouette

> Decide which of these objects is opaque and will block light. How could you test your ideas?

Shine a torch against a wall. Can you see the light on the wall? Put your hand in the torch beam. What can you see now? Your hand is opaque, it has blocked the light and **cast** a shadow. Try making an animal shadow using your hands.

Silhouettes

Artists have used shadows to make pictures and portraits. These pictures have no features. They are just dark shapes and like a shadow. **Silhouette** portraits were very popular two hundred years ago.

A silhouette portrait

不透明物体

一些材料会阻挡光。当光被阻挡时，你就能看到影子。不能让光通过的物体或材料是**不透明的**。光不能通过不透明的材料或物体。物体周围的光没有被阻挡。

单词表

cast 投射
opaque 不透明的
silhouette 轮廓；轮廓画

 请指出这些物体中哪些是不透明的且会阻挡光的？你要如何验证你的观点？

对着一面墙打开手电筒。你能观察到墙上的光吗？把你的手放到手电筒的光束里。现在你能观察到什么？你的手是不透明的，它阻挡了光的传播，从而**投射**出了影子。你可以尝试用双手摆出一个动物的形状。

轮廓

艺术家们利用影子来绘画和创作肖像画。这样的画没有具体的容貌特征，仅仅只有黑色的形状，像影子一样。**轮廓**肖像画在两百年前就已经十分流行了。

一幅轮廓肖像画

18

Make a silhouette gallery

You will need:

black paper   strong light source   chalk   scissors

1. Pin a piece of black paper onto a wall.

2. Place a friend, looking to the side, between the paper and a strong light source.

3. Draw around the shadow of their head cast on the paper.

4. Cut out the shape and mount it in your gallery.

Can you guess who is who?

制作一个轮廓画画廊

你将需要:

黑纸　　　　　强光源　　　　　粉笔　　　　　剪刀

1 将一张黑纸钉在墙上。

2 让你的朋友站在黑纸前,面向一侧,你的朋友在强光源和黑纸中间。

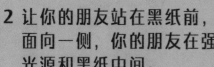

3 用笔画出你的朋友在黑纸上投射出的影子的轮廓。

4 将轮廓剪出来,贴到你的画廊里。

你能猜出轮廓对应的是谁吗?

Science Skills

What material blocks most light? Test it!

Word Box
scattered
translucent
transparent

Light cannot pass through opaque materials. The shadows cast by opaque objects are dark with sharp edges. Not all materials are opaque. Some materials let almost all light through. They cast almost no shadow. We call these materials **transparent**. The glass in windows is transparent; it allows almost all light through.
There are some materials that are **translucent**. Translucent materials let some light through but some of the light is **scattered**. The shadows they cast are pale and blurred.

Window glass lets light pass through it. It is transparent.

The best blind

The children wanted to make a blind for their sunny classroom. They needed a material that would stop the sun getting in their eyes but let enough light through so they could see. They decided they needed to know which materials were opaque, translucent and transparent.

Yes but we don't want to be in the dark.

We need something to block this light. It's hurting my eyes.

Talk to your partner. Define the new words you have learned 'translucent, opaque and transparent'. Do you both agree? Check the meaning in the glossary.

科 学 技 能

> 什么材料阻挡的光最多?
> 做个实验吧!

单词表

scattered 散射的
translucent 半透明的
transparent 透明的

光不能通过不透明的材料传播。不透明物体形成的影子是深色的且边缘分明,但并不是所有的材料都是不透明的。有些材料几乎能透过全部的光,不会形成影子。我们把这样的材料称作**透明的**。窗户上的玻璃是透明的,它几乎能透过所有的光。

还有一些材料是**半透明的**,半透明的材料能让部分的光通过,而让另一部分的光发生**散射**。半透明物体投射出的影子是比较暗淡模糊的。

窗户上的玻璃能让光通过,它是透明的。

是的,但是我们也不想身处黑暗之中。

我们需要用一样东西来挡住光,这光太晃眼睛了。

最好的百叶窗

小朋友们想在阳光明媚的教室里安装一个百叶窗。他们需要一种材料,这种材料既能挡住大部分的光而不晃眼睛,又能让他们在教室里看得见。他们决定弄清楚哪些材料是不透明的、半透明的和透明的。

> 和你的小伙伴讨论一下。定义刚刚学到的三个新词语:半透明的、不透明的和透明的。你们的观点一致吗?对照后面的第三部分科学词汇加油站的释义检查一下吧。

The children had different ideas about what to do.

Make a plan of your own. Think about
What will you need?
What will you do?
What will you change?
What will you measure or observe?
How will you record what you find out?
How will you decide what your **evidence** shows?

小朋友们对于要如何做有不同的观点。

单词表
evidence 证据

自己制订一个方案。思考一下：

你需要什么材料？

你要怎么做？

你要改变什么？

你要测量或观察什么？

你要如何记录你的发现？

你要如何**根据**证据做出决定？

Science Skills

How can shadows change? Test it!

Shadows do not always look the same. Some look very dark, others are lighter. Some have sharp, crisp edges, others are **blurred**.
The size of shadows can also change.
The children were making a shadow puppet play about sailing to a desert island. They wanted the island to look bigger as their sail boat got closer to it. How can they do this?

Word Box
blurred

> If we make lots of different sized islands each one getting bigger, we can keep changing them and it will look as if the island is getting closer.

> Everyone will see us doing that though. It won't look good.

> Let's see how many ways we can change the shadow. I'm just going to use a simple shape to try things out.

> We could change the material. Would that work?

> Make a simple shadow puppet. Explore how you can make the shadow change.
> You will need: paper, card, scissors, sticky tape, lolly stick or straw, lamp or torch, tape measure...
> Spend some time exploring how to change the shape of the shadows. How can you make the shadows bigger or smaller? What else could you move?

科 学 技 能

影子是如何改变的？做个实验吧！

影子并不是完全一样的。一些影子比较暗，而另外一些影子比较亮。一些影子的边缘轮廓分明，而另外一些影子的边缘轮廓是**模糊的**。

影子的大小同样也可以改变。

小朋友们在制作一个荒岛漂流的皮影戏。当船离荒岛更近时，他们想让荒岛看上去更大。这要如何实现呢？

> 单词表
> blurred 模糊的

> 如果我们制作出很多大小不同的荒岛，然后从小到大依次更换，这样看上去似乎船离岛就越来越近了。

> 这样大家都会看到我们更换，表演的效果不好。

> 让我们来思考一下改变影子大小有多少种方法。我想用一个简单的方法尝试一下。

> 我们可以改变材料。这个方法可以吗？

制作一个简单的皮影，探索一下你要如何改变影子的大小。

你将需要：纸张、卡片、剪刀、胶带、木棍或吸管、台灯或手电筒、卷尺……

花点儿时间探索一下如何改变影子的形状。你要如何才能让影子变大或变小？还有什么是你可以移动的？

The children experimented with a simple shape. They drew around each shadow and measured the height. Then they moved the shape further away. They drew around the shadow again and measured the height. They repeated this with the shape further away from the light.

They drew a table of their results.

Distance from light (cm)	Height of shadow (cm)
10	39
20	33
30	31
40	30
50	29
60	28

Try it yourself or use their results.

1 What is the pattern between the distance of the puppet from the light source and the size of the shadow?
2 Draw a picture to explain why you think the shadows changed like this.
3 Write a rule about what you have found out.

小朋友们用一个形状简单的物体进行了实验。他们绕着影子画出了形状并测量了高度。然后他们将物体移得远一点儿。他们再次画出形状并测量高度。然后他们再移得更远，并重复了上述操作。

他们将结果制作成了表格。

距离光源的距离（厘米）	影子的高度（厘米）
10	39
20	33
30	31
40	30
50	29
60	28

自己尝试一下或者使用表格中的结果。

1 皮影距离光源的远近和影子的大小之间有什么规律？
2 画图解释一下为什么你认为影子如此变化的原因。
3 将你的发现用一条规律总结出来。

The Sun Is Powerful

The Sun is an important and powerful source of light. Without it the world would be cold and in darkness.

The Sun gives out heat and light.

1 What would our world be like without the Sun?

The Sun gives us light and it also gives us warmth. Even though the Sun is millions of kilometres away from us we can still feel its heat. However, we need to be careful: the Sun can be dangerous.

Our skin can burn if we stay in the Sun for too long or go out when the Sun is very hot. We can damage our eyes by looking at the Sun. We must never look at the Sun directly, even if we are wearing sunglasses.

2 Where is the sunniest place you have been?

Wearing a sun hat, putting protective sun cream on our skin and staying in the shade on very sunny days are things we can do to stop our skin being damaged by the Sun.

If we wear loose clothing and drink lots of water on sunny days, we are more likely to stay cool and not get too hot or too thirsty as our bodies lose water in the heat.

📓 Make a poster warning of the dangers of too much Sun.

太阳是十分强大的

太阳是一个重要且强大的光源。如果没有太阳，我们的世界就会陷入寒冷和黑暗。

太阳发出热和光。

1 如果没有太阳，我们的世界会变得怎样？

太阳给予我们光和温暖。即使它离我们数百万公里，但我们仍然能感受到太阳的热量。但是，我们也要小心：因为太阳有时会很危险。

如果我们在太阳下停留太久，或者在太阳光很强的时候外出，皮肤可能会被晒伤。当我们直视太阳时，眼睛可能会受伤，所以一定不要直视太阳，即使我们戴着墨镜也不行。

2 你去过太阳光最强烈的地方是哪里？

为了避免皮肤被太阳晒伤，我们可以戴防晒帽，把防晒霜抹到皮肤上或者待在阴凉处。

由于我们的身体在炎热的环境中会流失水分，所以在晴天我们可以身着宽松的衣物、多喝水，这样可以保持凉爽，不会感觉太热或太渴。

📖 制作一张海报，警告人们被太阳曝晒的危害。

Natural Light

The Sun is a **natural** source of light. It is not human-made like torches or street lights.

> **Word Box**
> natural

Sort these light sources into natural and human-made.

Light from animals

Some animals live in very dark places like caves or in the deep ocean where there is not much light. Animals like these can make their own light. They are their own sources of light. The firefly has a tail that can light up.

In Japan fireflies used to be caught, put in small cages and used as lights.

The light on the angler fish attracts food.

Look at this amazing fish. It's called an angler fish and lives in the deep, dark ocean. It makes its own light to attract smaller fish so it can eat them!

自然光

太阳是一种**自然**光源。它和像手电筒或路灯那样的人造光源不一样。

单词表
natural 自然的

把下图的光源分为自然光源和人造光源。

动物发出的光

有些动物生活在洞穴或深海这种没有太多光的地方。这些动物可以自己发光,它们自身就是光源。萤火虫的尾部就能发光。

在日本,以前人们会将萤火虫捉起来,放进小笼子里用作灯。

安康鱼通过发光来吸引猎物。

观察这条神奇的鱼,它叫作安康鱼,生活在黑暗的深海里。安康鱼可以发光,以吸引小鱼过来,从而可以捕食它们!

What I Know about Light and Dark

Complete this spider diagram about what you already know about light and dark. Here are some words to help you.

    source dark light absence Sun see
    shadow night day reflect

Light and dark

Sources of Light

Something that makes light is called a light **source**.

1 Write the name of each object in the correct column.

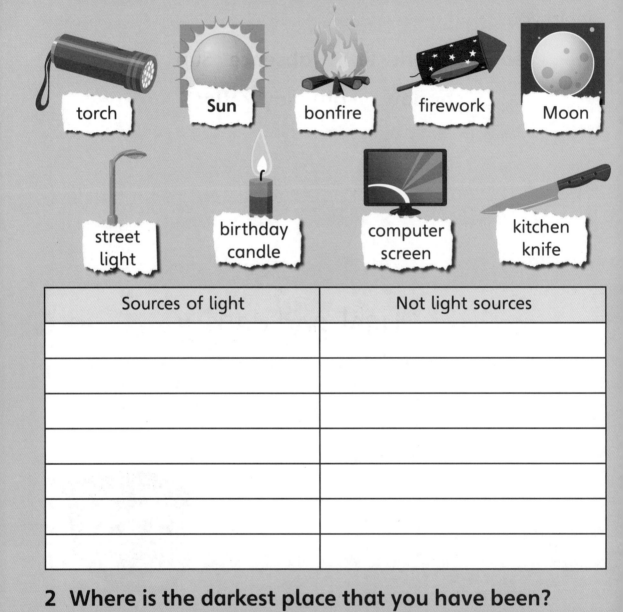

Sources of light	Not light sources

2 Where is the darkest place that you have been?

Bright and Dim

We need light to see.
Where there is no light we say it is dark.

1. Complete the table.
2. Circle the brightest and darkest places in your list.

Dark places	Light places
Under the bed	On the playground

Not all lights are the same. Some are very bright, others are quite dim.

3. Put these light sources in the right order. Put the brightest as number 1 and the dimmest as number 4.

football stadium floodlights

child's nightlight

car headlamp

table lamp

Brightest
1 _____ 2 _____
3 _____ 4 _____
Dimmest

Reflecting Light

Some objects do not make their own light but reflect it well. We call these objects **reflectors** of light.

1. Draw and label two good reflectors of light.

2. Write the names of the objects from the picture in page 7 on the Prat one in the correct group. One has been done for you.

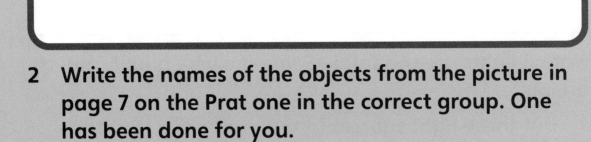

light sources: street lamp

light reflectors:

Add two more objects to each group.

3. What do good reflectors all have in common?

What Is a Shadow?

1 Look at the picture. Why does the palm tree cast a shadow?

2 What is the source of light in the picture?

3 Explain why the shadow of the palm tree is similar in shape to the real tree.

Draw yourself outside on a sunny day.
Put the Sun into your picture.
Draw your shadow in the picture.

Making Shadows

Light travels in straight lines. It cannot travel around objects. Some things block light and **cast** shadows.

1   Link the silhouettes to the correct creature. One has been done for you.

2   Draw a different animal. Make it into a silhouette.

Show it to a friend. Can your friend guess what your silhouette animal is?

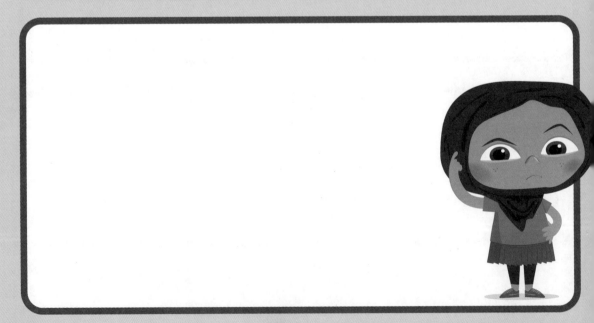

Opaque Objects

Some objects are **opaque**. They do not let light through.

1. Circle the opaque objects.

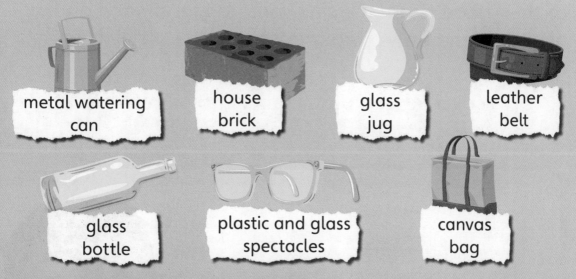

metal watering can   house brick   glass jug   leather belt

glass bottle   plastic and glass spectacles   canvas bag

Opaque objects can cast shadows.

2. Label this shadow diagram. Use these words to help you.

    light source   shadow   wall

    light ray   beam of light

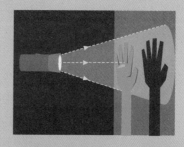

3. Now complete and label this shadow diagram.

4. Draw the shadow cast on the wall by the airflow ball.

Making a Silhouette Portrait

Children are making silhouette portraits. They use a bright torch as a light source. Children take turns to cast a shadow of their head (from the side) onto paper pinned onto a wall. A friend draws around the shadow. They cut out the shape.

1 **Draw and label a diagram to show how the shadow is made.**

2 **Draw in the path of light.**

3 **Write an explanation of how a silhouette shadow is made here.**

Science Skills

What material blocks most light? Test it!

1. Draw lines to match the word to the definition.

Opaque	A material that lets almost all light pass through
Translucent	A material that lets some light through
Transparent	A material that does not let any light pass through

2. Label these objects as translucent, transparent or opaque.

3. Draw and label three more objects. Choose one opaque, one transparent and one translucent object.

What materials block light? Test it!
Testing materials: Our investigation

We are trying to find out

We are changing

We are measuring or observing

Make a note of your results here.

What do your results show?

More about Materials

Different types of materials are useful for different jobs. Use your research skills to find out more about transparent, translucent and opaque materials.

Complete the table. One has been done for you.

Type of material	Examples	Uses
Transparent	Contact lenses	Let light pass through lens directly to the eye
Translucent		
Opaque		

Try this yourself

1. Investigate the types of shadows that each material produces.
2. Do all three types of materials produce shadows?
3. How would you describe the shadows made by each material, are they dark, sharp-edged, pale or fuzzy?

Fill in the table.

Type of material	Material I used	What sort of shadow is cast?
Transparent		
Translucent		
Opaque		

Science Skills

How can shadows change? Test it!

Children's investigation: Changing shadows

Children investigated how they could change the size of a shadow puppet. They made a simple shadow puppet shape and gradually moved the puppet away from the light.

They drew around the shadow the puppet made each time and measured the height.

Use their results to answer the question.

Here are their results:

Distance from light (cm)	Height of shadow (cm)
10	39
20	33
30	31
40	30
50	29
60	28

What questions can the results answer? For example, How far from the light was the puppet when the shadow was 30 cm tall?

Design a Shadow Puppet

Use this page to design your own shadow puppet. Think about where you want to cast shadows and where you might like light to pass through.

The Sun Is Powerful

The Sun is an important source of light but we need to take care as it can be harmful to us.

1. Write why these are good ideas or bad ideas. Explain your reasons.

   a Using protective sun cream on a sunny day is a good / bad idea because

   b Never wearing a sun hat is a good / bad idea because

   c Going out in the hottest part of the day is a good / bad idea because

   d Looking directly at the Sun is a good / bad idea because

Now write an idea of your own like the ones shown.

Design a 'Sun Safety' poster.

What I Know about Light and Dark

Show what you know about the topics in the boxes below. You could write a sentence, draw a picture or diagram or do all of these things.

Sources of light	How light travels

How shadows form	How to change the size of a shadow

Glossary

absence: lack of something

beam: the light from a light source in any particular direction e.g. the beam from a torch

blurred: not clear, fuzzy

cast: to make a shadow appear on a surface

dark: little or no light

evidence: something that shows that something else is true

light: something that lets us see objects

light ray: lines in a diagram showing the direction light travels

natural: found in nature, not human-made or artificial

opaque: materials that do not let light pass through

reflect: when light bounces off a surface

reflection: what we see in a mirror, or on a shiny surface

reflector: an object or material that reflects light well

scattered: move quickly in different directions

shadow: a dark shape on a surface made by opaque objects blocking light

silhouette: the dark shape of something against a lighter background

source: where something comes from

straight: not bent or curvy, going in one direction

Sun: the Earth's nearest star and most important source of light

translucent: a material that lets some light through but scatters it or changes it in some way

transparent: a material that lets light pass through

词汇表

缺失：缺少某件事物

光束：光源朝某个特定方向发出的光。例如，手电筒发出的光束

模糊的：不清楚的，不清晰的

投射：让影子出现在物体的表面

黑暗的：很少的光或没有光

证据：能表明某件事物是真实的

光：能让我们看见物体的一种物质

光线：在一幅图中展示光传播方向的直线

自然的：自然界中本身存在的，非人造的或非人工的

不透明物：光无法穿过的材料

反射：光在物体表面发生反弹的现象

镜像：在平面镜中或光滑表面上看到的映像

反光体：能很好地反光的物体或材料

散射的：很快地在不同方向上分散开来

影子：不透明物体阻挡光线传播而在其他的表面形成的深色形状

轮廓：物体在更亮的背景下形成的深色的形状

来源：事物产生的地方

直线：不是弯曲的或曲形的，而是沿同一个方向的

太阳：离地球最近的恒星，地球上最重要的光源

半透明物体：能透过部分光线且以某种方式将光线散射或改变的材料

透明物体：能透过所有光线的材料

培生科学虫双语百科
奇妙物理

Changes of State
物态的变化

英国培生教育出版集团 著·绘
徐 昂 译

电子工业出版社
Publishing House of Electronics Industry
北京·BEIJING

Original edition, entitled SCIENCE BUG and the title Changes of State Topic Book, by Debbie Eccles published by Pearson Education Limited © Pearson Education Limited 2018
ISBN: 9780435195489

All rights reserved. No part of this book may be reproduced or transmitted in any form or by any means, electronic or mechanical, including photocopying, recording or by any information storage retrieval system, without permission from Pearson Education Limited.

This adaptation of SCIENCE BUG is published by arrangement with Pearson Education Limited.
Chinese Simplified Characters and English language (Bi-lingual form) edition published by PUBLISHING HOUSE OF ELECTRONICS INDUSTRY, Copyright © 2023.
For sale and distribution in the mainland of China exclusively (except Hong Kong SAR, Macau SAR and Taiwan).

本书中英双语版由Pearson Education（培生教育出版集团）授权电子工业出版社在中华人民共和国境内（不包括香港、澳门特别行政区及台湾地区）独家出版发行。未经出版者书面许可，不得以任何方式抄袭、复制或节录本书中的任何部分。

本套书封底贴有Pearson Education（培生教育出版集团）激光防伪标签，无标签者不得销售。

版权贸易合同登记号　图字：01-2022-2381

图书在版编目（CIP）数据

培生科学虫双语百科. 奇妙物理. 物态的变化：英汉对照 / 英国培生教育出版集团著、绘；徐昂译. --北京：电子工业出版社，2024.1
ISBN 978-7-121-45132-4

Ⅰ. ①培… Ⅱ. ①英… ②徐… Ⅲ. ①科学知识－少儿读物－英、汉 ②物质－状态－变化－少儿读物－英、汉 Ⅳ. ①Z228.1 ②O414.12-49

中国国家版本馆CIP数据核字（2023）第034690号

责任编辑：李黎明　文字编辑：王佳宇
印　　刷：河北迅捷佳彩印刷有限公司
装　　订：河北迅捷佳彩印刷有限公司
出版发行：电子工业出版社
　　　　　北京市海淀区万寿路173信箱　邮编：100036
开　　本：787×1092　1/16　印张：35　字数：840千字
版　　次：2024年1月第1版
印　　次：2024年2月第2次印刷
定　　价：199.00元（全9册）

凡所购买电子工业出版社图书有缺损问题，请向购买书店调换。若书店售缺，请与本社发行部联系，联系及邮购电话：（010）88254888，88258888。
质量投诉请发邮件至zlts@phei.com.cn，盗版侵权举报请发邮件至dbqq@phei.com.cn。
本书咨询联系方式：010-88254417，lilm@phei.com.cn。

使用说明

欢迎来到少年智双语馆！《培生科学虫双语百科》是一套知识全面、妙趣横生的儿童科普丛书，由英国培生教育出版集团组织英国中小学科学教师和教研专家团队编写，根据英国国家课程标准精心设计，可准确对标国内义务教育科学课程标准（2022年版）。丛书涉及物理、化学、生物、地理等学科，主要面向小学1~6年级，能够点燃孩子对科学知识和大千世界的好奇心，激发孩子丰富的想象力。

本书主要内容是小学阶段孩子需要掌握的物理知识，含9个分册，每个分册围绕一个主题进行讲解和练习。每个分册分为三章。第一章是"科学虫趣味课堂"，这一章将为孩子介绍科学知识，培养科学技能，不仅包含单词表、问题和反思模块，还收录了多种有趣、易操作的科学实验和动手活动，有利于培养孩子的科学思维。第二章是"科学虫大闯关"，这一章是根据第一章的知识点设置的学习任务和拓展练习，能够帮助孩子及时巩固知识点，准确评估自己对知识的掌握程度。第三章是"科学词汇加油站"，这一章将全书涉及的重点科学词汇进行了梳理和总结，方便孩子理解和记忆科学词汇。

2024年，《培生科学虫双语百科》系列双语版由我社首次引进出版。为了帮助青少年读者进行高效的独立阅读，并方便家长进行阅读指导或亲子共读，我们为本书设置了以下内容。

（1）每个分册第一部分的英语原文（奇数页）后均配有对应的译文（偶数页），跨页部分除外。读者既可以进行汉英对照阅读，也可以进行单语种独立阅读。问题前面的 🔖 符号表示该问题可在第二部分预留的位置作答。

（2）每个分册第二部分的电子版译文可在目录页扫码获取。

（3）本书还配有英音朗读音频和科学活动双语视频，也可在目录页扫码获取。

最后，祝愿每位读者都能够享受双语阅读，在汲取科学知识的同时，看见更大的世界，成为更好的自己！

电子工业出版社青少年教育分社
2024年1月

Contents 目录

Part 1　科学虫趣味课堂　　　　　　　　　　　/ 1

Part 2　科学虫大闯关　　　　　　　　　　　　/ 35

Part 3　科学词汇加油站　　　　　　　　　　　/ 51

Part2译文

配套音视频

What Do We Know about Solids, Liquids and Gases?

We are always surrounded by different materials. Materials can have a range of **properties**. A property describes something about a material. At room temperature some materials are **solid**, some **liquid** and others are a **gas**.

Word Box
gas
liquid
property
solid

- 1a Name a material that is a solid.
- 1b How do you know it is a solid?
- 2a Name a material that is a liquid.
- 2b How do you know it is a liquid?
- 3a Name a material that is a gas.
- 3b How do you know it is a gas?

Ajay has left his ice lolly in a warm kitchen.

- 4a What will happen to the ice lolly if he leaves it there all day?
- 4b Explain why.
- 5 With a partner, discuss things at home that change from a solid to a liquid, and from a liquid to a solid.

关于固体、液体和气体，我们知道什么？

我们被各种各样的物质包围，物质拥有各种**特性**，这些特性描述了物质具有的某种性质。在室温下，物质可能是**固体**、**液体**或**气体**。

单词表
gas 气体
liquid 液体
property 特性
solid 固体

📄 1a 说出一种固体物质。
📄 1b 你是如何知道它是固体的？
📄 2a 说出一种液体物质。
📄 2b 你是如何知道它是液体的？
📄 3a 说出一种气体物质。
📄 3b 你是如何知道它是气体的？

阿杰伊把他的冰棒放在了温暖的厨房。

📄 4a 如果他把冰棒放在那儿一整天，会发生什么？
📄 4b 解释原因。
📄 5 和你的小伙伴一起，讨论一下，家里有哪些物品可以从固体变为液体？哪些物品可以从液体变为固体？

Science Skills

Observe it!

Different types of materials may have some similar properties.

One way of sorting materials is to sort them into two groups. One group will share a property, for example they are all liquid. The other group will not have this property. We call this classifying.

One way of classifying the materials in the picture is grouping all the transparent materials together and all the materials that are not transparent together.

Which material?

You will need: the materials shown in the diagram, a hand lens...

1 Work in a small group. Explore and describe the properties of the materials.

2 Take it in turns to describe a property. Ask children in other groups to guess the material being described.

1 List different ways of classifying these materials into two groups.

Some materials keep a fixed shape. We say these are solid.

2 List the solid materials in the diagram.

科学技能

观察一下吧!

不同种类的物质可能拥有相似的特性。

一种分类方法是将物质分为两组。一组物质都有相同的特性,例如,都是液体。另一组则不具有这种特性。我们把这种方式叫作分类。

还有一种分类方法是将图中的物质按透明的和不透明的分为两组。

哪种物质?

你将需要:图中所示的物质、一个放大镜……

1 和你的小伙伴一起,探索并描述这些物质的特性。

2 轮流描述一种特性,让其他的小朋友猜一猜描述的物质是什么。

一些物质的形状固定。我们将这些物质叫作固体。

📓 1 列出能将这些物质分为两组的不同的分类方法。

📓 2 列出图中的固体物质。

Liquids and Solids

Orange juice is a liquid. When it is poured it takes the shape of the glass and has a flat top.

Pouring bubble bath

You will need: a bottle of bubble bath, a transparent plastic cup...

1 Work in a small group. Pour bubble bath from the bottle into the cup. Observe closely what happens.

2 Is bubble bath a liquid? Why?

3 How is it different from orange juice?

1a List some other liquids.

1b How are all liquids the same?

1c Discuss with a partner how they are different.

The salt is in very small pieces. We say it is granular. When it is poured it forms peaks.

Pouring salt

You will need: a bottle of salt, a transparent plastic cup...

1 Work in a small group. Pour salt from the bottle into the cup. Observe closely what happens.

2 Is salt a liquid? Why?

3 How is it different from milk?

2 List some other solids that we usually find in very small pieces.

液体和固体

橙汁是一种液体。当我们将橙汁倒进杯子里时，橙汁的形状和玻璃杯的形状相同，而且橙汁的顶部是平整的。

倾倒泡沫剂

你将需要：一瓶泡沫剂、一个透明的塑料杯……

1. 和你的小伙伴一起，将泡沫剂从瓶子里挤到杯子里。仔细观察发生了什么。
2. 泡沫剂是液体吗？为什么？
3. 泡沫剂与橙汁有什么不同？

📖 **1a** 列出一些其他的液体。

1b 所有的液体有哪些共同点？

1c 和你的小伙伴一起，讨论一下，这些液体有哪些不同。

盐由很小的颗粒组成。我们说盐具有颗粒状结构。当我们倾倒盐时，盐堆顶部呈峰状。

倾倒盐

你将需要：一瓶盐、一个透明的塑料杯……

1. 和你的小伙伴一起，将盐从瓶子里倒进杯子里。仔细观察发生了什么。
2. 盐是液体吗？为什么？
3. 盐与牛奶有什么不同？

📖 **2** 列出常见的其他小块或颗粒状的固体。

States of Matter

1 Look around your classroom. How many different materials can you name?

Word Box
matter
state

Everything around us is made of **matter**. Matter is the scientific word for stuff. The air we breathe is matter even though we cannot see it. Scientists classify matter into three types: solids, liquids and gases. We call these **states**.

2 Which state of matter is each of the different materials inside the bottles?

Air   Orange Juice   Rice

Empty bottle?

You will need: an empty plastic water bottle.

1 Work in a pair. Make sure the top is tightly screwed on the bottle. Squeeze the bottle.

2 What happens? Why?

3 Take the top off the bottle. Squeeze the bottle again. What happens now? Why?

物质的形态

1 环顾你的教室，你能说出多少种不同物质的名字？

单词表
matter 物质
state 形态

我们周围的一切都是由物质构成的。**物质**是"东西"的科学术语。尽管我们看不见，我们呼吸的空气也是物质。科学家将物质分为三种：固体、液体和气体。我们把这三种叫作物质的**形态**。

2 瓶子里的不同物质分别是哪种形态？

空气　橙汁　大米

空瓶子？

你将需要：一个空的塑料瓶。

1 和你的小伙伴一起，确保瓶盖拧紧。挤压一下瓶子。

2 发生了什么？为什么？

3 将瓶盖拿开，再挤压一下瓶子。发生了什么？为什么？

These objects are made from materials in different states of matter.

Materials are made up of many millions of particles. A particle is a very little bit of the material. It is so small that one particle is invisible to us.

The particles in solids are tightly packed together. The particles in liquids are touching but can move around each other. The particles in gases are much further apart and do not often touch.

> 3 Point to the pictures that:
> a show only a solid material
> b show only a liquid material
> c show a solid and a gas
> d show a solid and a liquid.

We can think of particles as looking like tiny balls.

> 4 If we could see the particles that make up a solid, liquid and gas, what would they look like? Draw ten particles for each state.

这些物体是由不同形态的物质构成的。

材料是由数以百万计的粒子构成的。一个粒子是物质中非常小的一部分。粒子足够微小，所以我们看不见。

固体中的粒子紧凑地排列在一起。液体中的粒子相互接触，而且可以相互流动。气体中的粒子相距较远，不相互接触。

我们可以把粒子想象成小球。

3 指出相应的图片：
  a 图片中展示的只有固体物质
  b 图片中展示的只有液体物质
  c 图片中展示的包括固体和气体
  d 图片中展示的包括固体和液体

4 如果我们能看见组成固体、液体和气体的粒子，那么这些粒子看上去是怎样的？为物质的每种形态画出十个粒子。

The Effect of Temperature

The **temperature** of something is how hot or cold it is. Heating or cooling a material changes its temperature and can also make other changes.

Word Box
frozen
melting
temperature

ice pops

You will need: a **frozen** ice pop and some warm hands!

1a Work with a partner. Take it in turns to hold the ice pop in warm hands. Observe closely what happens.

1b What change did you observe? Why?

2a Put the ice pop in a sunny place.

2b Discuss with your partner what you think will happen and why?

3 What do you think would happen if you put the ice pop back in the freezer?

Melting happens when a solid warms up enough to become a liquid.

Food that is often found in a freezer or fridge.

📖 What do you think will happen to each of these foods if they are left in a warm place for an hour?

温度的影响

物体的**温度**用来表示它是冷的还是热的。将物质加热或冷却会改变它的温度,也可能引起其他变化。

单词表
frozen 冷冻的
melting 熔化的
temperature 温度

冰棒

你将需要:一个冷冻的冰棒和温暖的双手!

1a 和你的小伙伴一起,轮流用温暖的手拿着冰棒。仔细观察发生了什么。

1b 你观察到了什么变化?为什么?

2a 将冰棒放到阳光充足的地方。

2b 和你的小伙伴一起,讨论一下,你认为会发生什么?为什么?

3 如果把冰棒再放回冰箱,你认为会发生什么?

当固体的温度上升得足够高时,固体会**融化**,变为液体。

经常放在冷冻柜或冰箱里的食物。

如果这些食物被放在一个温暖的地方一个小时,你认为会发生什么?

Changing Chocolate

Chocolate changes when we heat it. Solid chocolate can melt and turn to liquid in your hands!

Melting chocolate

You will need: a cup of hot water (not boiling), a saucer, two lolly sticks, two cupcake cases, a piece of milk chocolate...

1. Work with a partner. Fill the cup to the top with hot water. Place the saucer on top of the cup and the piece of chocolate on top of the saucer.
2. What do you think will happen to the chocolate? Why?
3. Leave the chocolate on top of the saucer for 10 minutes.
4. Use a lolly stick to scoop up chocolate and make chocolate blobs on each cupcake case.
5. What do you think will happen to the chocolate blobs after an hour? Why?

We can make chocolates of different shapes using special moulds.

- Record instructions on how to make different-shaped chocolates.

变化的巧克力

当我们加热巧克力时,它会发生改变。固体巧克力会熔化并在你的手中变成液体!

熔化的巧克力

你将需要:一杯热水(不是沸水)、一个茶碟、两根木棍、两个蛋糕杯、一块牛奶巧克力……

1 和你的小伙伴一起,将热水装满杯子。将茶碟放在杯子上面,再将一块巧克力放在茶碟上。

2 你认为巧克力会发生什么?为什么?

3 将巧克力放在茶碟上十分钟。

4 用一根木棍搅动巧克力,并在每个蛋糕杯中制作巧克力团。

5 你认为一小时后巧克力团会发生什么?为什么?

我们可以用特殊的模具将巧克力制作成不同的形状。

📔 记录一下制作不同形状的巧克力的说明。

Ice

Heating and cooling materials can change their properties. Freezing (or **solidifying**) happens when a liquid cools down enough to become a solid.

> **Word Box**
> solidify

When liquid water becomes very cold it turns into solid water. We call solid water, ice.

1a What does this picture show?
1b What was the weather like on the day the picture was taken?
1c How do you know?
1d What will happen on a warm day?
1e Why?

2 Why do they have to re-build the Icehotel?

3 How is it re-built? Ask your teacher for reference materials about the Icehotel and design a web page to advertise it.

The Icehotel in Sweden is the world's biggest igloo! It is made entirely of ice. They have to re-build it every year.

冰

对物质进行加热或冷却可以改变它们的特性。当液体冷却到一定程度变成固体时,这种现象叫作凝固。

当液态水处于非常冷的状态时,它就会变为固态水。我们把处于固态的水称为冰。

单词表

solidify 凝固

1a 这张照片展示了什么?
1b 这张照片中的天气如何?
1c 你是怎么知道的?
1d 在温暖的天气里会发生什么?
1e 为什么?

2 为什么他们要重建冰旅馆?

3 冰旅馆是如何被重建的?向你的老师询问关于冰旅馆的参考资料,并设计一个网页去宣传它。

瑞典的冰旅馆是世界上最大的冰屋!它是完全由冰建成的,人们每年都要重建它。

Science Skills

Measure it!

We measure temperature using a thermometer.

Different materials change state at different temperatures.

All of these materials are shown in their solid state.

A

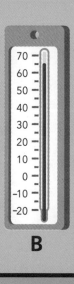
B

C

D

E

- 1a Read the temperatures on these thermometers.
- 1b Predict which material will melt at each temperature.
- 2 Use reference materials to check your predictions.

科学技能

测量一下吧！

我们用温度计来测量温度。

在不同的温度下，不同物质的形态会发生改变。

图中所示的这些物质都是固态的。

 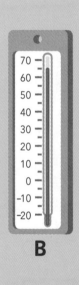   

　A　　　　B　　　　C　　　　D　　　　E

- 1a 读取温度计上的温度数值。
- 1b 请预测一下，哪种材料会在上图所示的温度计的温度下熔化。
- 2 使用参考资料来检验你的预测。

Science Skills

We measure temperature using a thermometer. There are lots of different types of thermometer.

  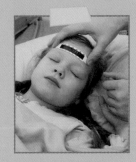 

Did you know?

Mercury is the only metal that is liquid at room temperature.

科学技能

我们使用温度计来测量温度。温度计有很多种类。

  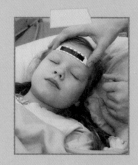 

你知道吗？

汞是唯一一种在室温下呈液态的金属。

Freezing and Melting Bathroom Liquids

We use different products in the bathroom to help us keep clean.

These bathroom products are liquids at room temperature.

Freezing bathroom products

You will need: shampoo, baby oil, hair conditioner, body lotion, bubble bath, mouthwash, six transparent plastic cups, an ice tray...

1 Work in a small group. Put a little of each bathroom product in a different cup.
2 Look at them and feel them. Describe them.
3 Place a little of each one in an ice tray compartment. Put the ice tray in the freezer and leave it overnight.
4 What do you think will happen to each of the bathroom products? Why?

If we take the frozen bathroom products out of the freezer and warm them up to room temperature they will melt.

凝固和熔化浴室里的液体

我们在浴室会使用不同的产品来保持清洁。

在室温下这些浴室里的产品是液态的。

凝固浴室里的产品

你将需要：洗发水、婴儿润肤油、护发素、身体乳、泡沫剂、漱口水、六个透明的塑料杯、一个制冰盘……

1. 和你的小伙伴一起，将浴室里的每一种产品取出一点儿放进不同的杯子里。
2. 观察并感受这些产品。描述一下。
3. 分别将每种产品放在制冰盘的小格子里。再把制冰盘放进冰箱，放置一整晚。
4. 你认为这些浴室里的产品会发生什么变化？为什么？

如果我们将凝固的浴室里的产品从冰箱里拿出来，放置在室温下，它们会熔化。

1 Which bathroom product will melt quickest?
2 Which bathroom product will melt slowest?
3 How could you investigate to find out?

Most materials are easy to identify as a solid, liquid or gas, but some are more unusual.

Paste and foam

You will need: a tube of toothpaste, a can of shaving foam, a plastic plate...

1 Work in a small group. Put a little toothpaste and shaving foam on the plate.
2 Decide as a group whether these are a solid, liquid or gas. Why?
3 What do you think would happen if you put them in the freezer overnight? Why?
4 Try it. Put them in a freezer overnight. What happened?

> 1 浴室里的哪种产品熔化得最快?
> 2 浴室里的哪种产品熔化得最慢?
> 📓 3 你是如何得到结果的?

大多数物质都很容易被确认为是固体、液体还是气体，但有一些物质则不太一样。

糊状物和泡沫

你将需要：一管牙膏、一罐剃须泡沫、一个塑料盘……

1. 和你的小伙伴一起，将少量的牙膏和剃须泡沫放到塑料盘中。
2. 和你的小伙伴一起，确定牙膏和泡沫究竟是固体、液体还是气体？为什么？
3. 如果将它们放进冰箱里一整晚，你认为会发生什么？为什么？
4. 尝试一下，将它们放进冰箱里，经过一整晚后，会发生什么？

Investigating Evaporation

Word Box
evaporation

1 Why are these wet clothes hanging on a clothes line outside?
2 What happens to them? Where does the water go?

When a liquid changes to a gas we call the process evaporation. Evaporation happens more quickly when it is hot or the air is moving.

3 Would the clothes dry as quickly inside the house? Why?

Observing evaporation

You will need: a green or blue paper towel, a cup of water, a paint brush...

1 Use the paint brush to write your name in water on the paper towel.
2 Leave the paper towel on your desk or table.
3 What do you think will happen to your name? Why?

We can see what happens to the water but it can be tricky to explain how it happens. Scientists use models to help them.

探究蒸发

单词表

evaporation 蒸发

当液体变为了气体，我们称这一过程为蒸发。

当温度较高或空气流通时，蒸发过程会加快。

1 为什么将这些湿衣服挂在外面的晾衣绳上？

2 它们发生了什么？水分去哪里了？

3 这些衣服在室内变干的速度会和在室外一样吗？为什么？

观察蒸发

你将需要：一条绿色或蓝色的纸毛巾、一杯水、一把漆刷……

1 用漆刷在水下的纸毛巾上写下你的名字。

2 将纸毛巾放到你的桌子上。

3 你认为你的名字会发生什么变化？为什么？

我们可以观察到水发生的变化，但可能很难解释这是如何发生的。科学家使用模型来进行解释。

In this model the balls represent the tiny particles the water is made from.

4 Which diagram shows water as a liquid? Describe how it is a model of a liquid.

5 Describe how the other diagram models water as a gas.

6 Talk to a partner. How is this a model of evaporation?

Balls

You will need: 12 hollow plastic balls, a large square piece of material...

1 Work in a group of five. Four of the group hold an edge of the square of material. Pull it tight.
  Ask your classmate to put the balls in the middle of the material.

2 Move the material slightly so the balls move on the material but do not jump off it.

3 Then move the material very quickly up and down so the balls jump off the material.

4 Then move your arms more slowly. What happens?

5 What do you think you have modelled?

在这个模型中，小球代表了构成水的微小粒子。

4 哪张图说明了水是液体？描述你认为这是液体模型的原因。
5 描述另一张图是水蒸气模型的原因。
6 和你的小伙伴一起，讨论一下，为什么这是蒸发的模型？

球

你将需要：12个空心塑料球、一大块方形布料……

1 以五个人为一个小组。其中四个人分别手持方形布料的一端，将布料拉紧。
让另外一个人把球放在布料的中间。
2 轻轻移动布料，让球在布料上移动但不会离开布料。
3 快速地上下移动布料，这样球就离开了布料。
4 然后慢慢地移动手臂。发生了什么？
5 你认为你们所建的模型代表了什么？

Investigating Condensation

Word Box
condensation
water vapour

When a gas cools and changes to a liquid, we call the process **condensation**. These pictures show examples of when condensation has occurred.

When water is a gas we call it **water vapour**. Water vapour is always in the air around us. We cannot see it until condensation occurs.

Breathe on a spoon

You will need: a cold, dry, shiny metal spoon...

1 Hold the back of the end of the spoon close to your mouth. Do not let it touch your mouth.

2 Breathe out hard onto the back of the spoon. Repeat three more times.

3 What has happened to the back of the spoon? Why?

4 Gently rub the back of the spoon with the tip of one finger. What do you feel? Why?

5 What does the spoon look like now you have rubbed it? Why?

探究液化

单词表
condensation 液化
water vapour 水蒸气

当气体冷却变为液体时，我们称这一过程为**液化**。这些照片展示了发生液化的例子。

当水处于气态时，我们称之为**水蒸气**。水蒸气一直存在于我们周围的空气中。直到液化发生以后，我们才能看到它。

对着勺子呼气

你将需要：一个冷的、干燥的、闪闪发光的金属勺子……

1. 将勺子末端的背面贴近嘴唇。不要让勺子接触嘴。
2. 用力对着勺子背面呼气。重复三次以上。
3. 勺子背面发生了什么变化？为什么？
4. 用指尖轻揉勺子背面，你感觉如何？为什么？
5. 在你轻揉勺子后，勺子看上去发生了什么变化？为什么？

The Water Cycle

On Earth we find water in all three states: water, ice and water vapour.

Word Box
water cycle

Why does water naturally change state on Earth?

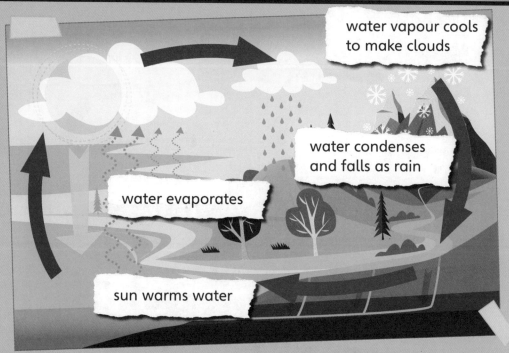

water vapour cools to make clouds

water condenses and falls as rain

water evaporates

sun warms water

In nature, water is constantly changing state from liquid to gas and back again. We call this the **water cycle**.

In the water cycle, water is warmed by the Sun. It evaporates from the sea, rivers and lakes. The water turns to vapour that rises in the air.

Water vapour high in the air becomes cold. It changes back to liquid and forms clouds.

As the clouds cool, the water droplets join together to form rain. Rain falls on the land and eventually runs back into the sea. The cycle starts again!

水循环

我们发现地球上的水有三种形态：水、冰和水蒸气。

单词表
water cycle 水循环

为什么地球上的水会自然地改变形态？

在自然界中，水会不断地从液态变为气态，再由气态变为液态。我们将这样的过程称为**水循环**。

在水循环中，水经太阳加热，从大海、河流和湖泊中蒸发。液态水会变为空气中的水蒸气。

水蒸气进入空气后，在高处遇冷会重新变为液态水，形成云。

随着云冷却，小水滴聚集在一起形成雨。雨落在大地上，最终流回大海。然后循环就会再次开始！

Did you know?

How old is your water? You might think it is fresh from the tap but because water is constantly recycled, it's almost as old as Earth itself!

When a liquid changes to a gas, we call the process 'evaporation'.

When a gas cools and changes to a liquid, we call the process 'condensation'.

Find out

Find out why garden ponds often freeze in winter but the sea rarely does.

Did you know?

98% of the world's water is liquid. Less than 2% is ice.

你知道吗?

你喝的水年龄是多大呢?你可能认为水龙头流出的水是新鲜的,但由于水一直处于循环之中,所以,水的年龄几乎和地球的年龄一样大!

当液体变为气体时,我们将这个过程称为"蒸发"。

当气体冷却变为液体时,我们将这个过程称为"液化"。

发现

试着去发现一下,为什么花园里的池塘总是在冬天结冰,而海水却很少结冰。

你知道吗?

世界上98%的水都是液态的。冰的占比小于2%。

What Do We Know about Solids, Liquids and Gases?

Everywhere we look there are different materials.

1a Name a material that is a solid.

_____

1b How do you know it is a solid?

_____

2a Name a material that is a liquid.

_____

2b How do you know it is a liquid?

_____

3a Name a material that is a gas.

_____

3b How do you know it is a gas?

_____

4 What will happen to an ice lolly left in a warm place? Explain your answer.

_____
_____
_____

A lolly is made out of ice.

Science Skills

Observe it!

These materials can be classified into a group that are soft and a group that are not soft.

1 List three other ways you could sort these materials.

Are _____ are not _____

Are _____ are not _____

Are _____ are not _____

2 List some solid and not solid materials from the diagram.

Solid	Not solid

Liquids and Solids

1 List four different liquids.

2 Choose the two liquids you think are the most different from each other. Explain how they are different.

3 List four granular solids.

4 Choose the two granular solids you think are the most different from each other. Explain how they are different.

Solid, Liquid or Gas?

Think about the properties of wood, water and air.

1. Are these statements true or false?

Solids have a fixed shape. _____

You can pour some solids. _____

Gases have a fixed shape. _____

If you pour a liquid it will have a flat top. _____

Air is all around us. _____

Liquids have a fixed shape. _____

All solids have the same properties. _____

There are different types of liquids. _____

There is only one type of gas. _____

2. How would you describe wood, water and air to a young child?

Wood is _____

Water is _____

Air is _____

We often think of particles as looking like small balls.

3. If we could see the particles that make up a solid, liquid and gas, what would they look like? Draw ten particles for each state. Explain your drawings.

solid

liquid

gas

Science Skills

Predict it!
Some solid materials melt when they are heated.

1a Identify the foods in these diagrams.

1b Predict which will melt if they are left in a warm place for an hour.

1c Clearly record your predictions below.

_____
_____
_____
_____

Changing Chocolate

Record instructions to make different-shaped chocolates using special moulds. You can use drawings to help make the instructions clear.

Ice

Ask your teacher for information on the Icehotel in Sweden. Design a web page to advertise it. Include information about how it is built and how nice it is to stay in.

Science Skills

Measure it!

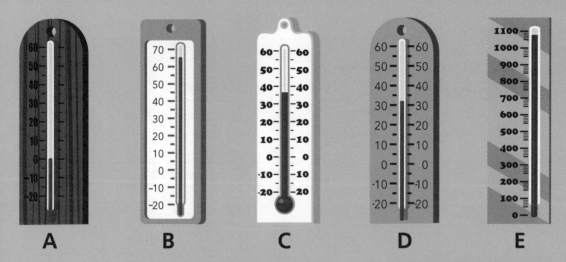

A   B   C   D   E

Each of these thermometers shows the temperature at which one of these materials melts: butter, wax, chocolate, ice, and copper.

Complete the table

1. Record the temperatures displayed on the thermometers.
2. Predict which material will melt at which temperature.
3. Research melting points of each of the materials and record the material that melts at each temperature.

Thermometer	Temperature	Type of material predicted to melt	Type of material actually melts
A			
B			
C			
D			
E			

Bathroom Products

1 Describe these bathroom products.

Shampoo _____

Hair conditioner _____

Baby oil _____

Body lotion _____

Bubble bath _____

Mouthwash _____

2 Which ones do you think will solidify in the freezer? Why?

_____
_____
_____
_____
_____
_____
_____
_____

Science Skills

Plan it!

This ice tray contains a frozen sample of shampoo, baby oil, hair conditioner, body lotion, bubble bath and mouthwash.

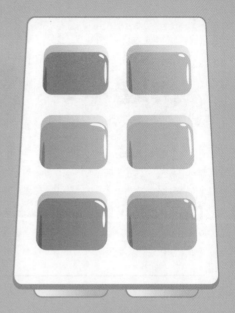

How could you find out how long it takes for the different products to melt at room temperature? Record your plan.

Investigating Evaporation

1. Describe how the first picture is a model of a liquid.

2. How does the other picture show water as a gas?

3. Explain evaporation. You can use a diagram to help.

4 What can we do to speed up evaporation?

_____

_____

5a List three pieces of equipment that can be used to speed up evaporation.

5b Give an example of when each one is used.

5c Explain why it speeds up evaporation.

Equipment 1 is _____

_____

_____

_____

Equipment 2 is _____

_____

_____

_____

Equipment 3 is _____

_____

_____

_____

Investigating Condensation

Gas cooling and becoming a liquid is called condensation.

Draw three pictures that show different times and places that condensation occurs and explain why.

	Condensation occurs because _____ _____ _____ _____
	Condensation occurs because _____ _____ _____ _____
	Condensation occurs because _____ _____ _____ _____

The Water Cycle

In nature, water is constantly changing state from liquid to gas and back again. We call this the water cycle.

Use this picture and what you have learned about evaporation and condensation to record five questions about the water cycle.

Q1 _____

Q2 _____

Q3 _____

Q4 _____

Q5 _____

What Do We Know about States of Matter?

1. What are the three states of matter?

   _____

2. Look at the diagram. It shows materials in all three states of matter. Identify and label the material in each of the three states.

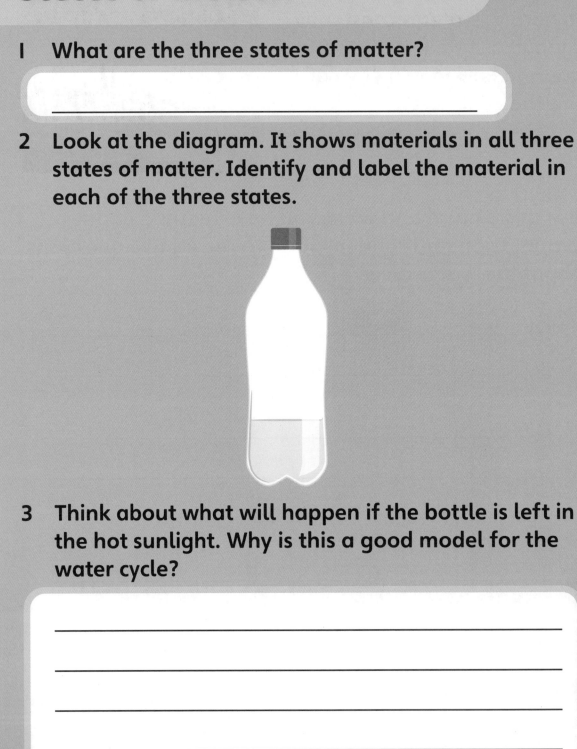

3. Think about what will happen if the bottle is left in the hot sunlight. Why is this a good model for the water cycle?

   _____
   _____
   _____
   _____

Glossary

condensation: gas cooling enough to form a liquid

evaporation: liquid warming up enough to become a gas

frozen: turned to a solid because of extreme cold

gas: a material that fills the space available; a gas is often invisible

liquid: a material that flows freely; in a container it takes the container's shape and settles with a flat top

matter: scientific word for stuff; it can exist as a solid, liquid or gas

melting: a solid warming up enough to become a liquid

property: describes something about a material

solid: a material with a fixed shape

solidify: liquid cooling enough to become a solid

state: the form a substance is in (solid, liquid or gas)

temperature: measure of how hot or cold something is

water cycle: water in nature changing from liquid to gas and back again

water vapour: particles of water that are always in the air around us

词汇表

液化：气体充分冷却变为液体的过程

蒸发：液体的温度升高变为气体的过程

冷冻的：由于十分寒冷而变成了固体

气体：一种充斥在空间里的物质，气体通常是看不见的

液体：一种能随意流动的物质，它的形状和容器的形状一致且顶部平坦

物质："东西"的科学术语，能以固体、液体、气体的形式存在

熔化：固体的温度升高变为液体的过程

特性：用于描述物质的性质

固体：有固定形状的物质

凝固：液体充分冷却变为固体的过程

形态：物质存在的形式（固体、液体、气体）

温度：衡量物体冷或热的方式

水循环：自然界中的水从液体变为气体，再由气体变为液体的过程

水蒸气：充斥于我们周围空气中的水分子

培生科学虫双语百科
奇妙物理

Electricity

电

英国培生教育出版集团 著·绘
徐 昂 译

电子工业出版社
Publishing House of Electronics Industry
北京·BEIJING

Original edition, entitled SCIENCE BUG and the title Electricity Topic Book, by Debbie Eccles published by Pearson Education Limited © Pearson Education Limited 2018
ISBN: 9780435195649

All rights reserved. No part of this book may be reproduced or transmitted in any form or by any means, electronic or mechanical, including photocopying, recording or by any information storage retrieval system, without permission from Pearson Education Limited.

This adaptation of SCIENCE BUG is published by arrangement with Pearson Education Limited.

Chinese Simplified Characters and English language (Bi-lingual form) edition published by PUBLISHING HOUSE OF ELECTRONICS INDUSTRY, Copyright © 2023.

For sale and distribution in the mainland of China exclusively (except Hong Kong SAR, Macau SAR and Taiwan).

本书中英双语版由Pearson Education（培生教育出版集团）授权电子工业出版社在中华人民共和国境内（不包括香港、澳门特别行政区及台湾地区）独家出版发行。未经出版者书面许可，不得以任何方式抄袭、复制或节录本书中的任何部分。

本套书封底贴有Pearson Education（培生教育出版集团）激光防伪标签，无标签者不得销售。

版权贸易合同登记号　图字：01-2022-2381

图书在版编目（CIP）数据

培生科学虫双语百科. 奇妙物理. 电：英汉对照 / 英国培生教育出版集团著、绘；徐昂译. --北京：电子工业出版社，2024.1
ISBN 978-7-121-45132-4

Ⅰ.①培… Ⅱ.①英… ②徐… Ⅲ.①科学知识－少儿读物－英、汉 ②电－少儿读物－英、汉 Ⅳ.①Z228.1 ②O441.1-49

中国国家版本馆CIP数据核字（2023）第035077号

责任编辑：李黎明　文字编辑：王佳宇
印　　刷：河北迅捷佳彩印刷有限公司
装　　订：河北迅捷佳彩印刷有限公司
出版发行：电子工业出版社
　　　　　北京市海淀区万寿路173信箱　邮编：100036
开　　本：787×1092　1/16　印张：35　字数：840千字
版　　次：2024年1月第1版
印　　次：2024年2月第2次印刷
定　　价：199.00元（全9册）

凡所购买电子工业出版社图书有缺损问题，请向购买书店调换。若书店售缺，请与本社发行部联系，联系及邮购电话：（010）88254888，88258888。
质量投诉请发邮件至zlts@phei.com.cn，盗版侵权举报请发邮件至dbqq@phei.com.cn。
本书咨询联系方式：010-88254417，lilm@phei.com.cn。

使用说明

欢迎来到少年智双语馆！《培生科学虫双语百科》是一套知识全面、妙趣横生的儿童科普丛书，由英国培生教育出版集团组织英国中小学科学教师和教研专家团队编写，根据英国国家课程标准精心设计，可准确对标国内义务教育科学课程标准（2022年版）。丛书涉及物理、化学、生物、地理等学科，主要面向小学1~6年级，能够点燃孩子对科学知识和大千世界的好奇心，激发孩子丰富的想象力。

本书主要内容是小学阶段孩子需要掌握的物理知识，含9个分册，每个分册围绕一个主题进行讲解和练习。每个分册分为三章。第一章是"科学虫趣味课堂"，这一章将为孩子介绍科学知识，培养科学技能，不仅包含单词表、问题和反思模块，还收录了多种有趣、易操作的科学实验和动手活动，有利于培养孩子的科学思维。第二章是"科学虫大闯关"，这一章是根据第一章的知识点设置的学习任务和拓展练习，能够帮助孩子及时巩固知识点，准确评估自己对知识的掌握程度。第三章是"科学词汇加油站"，这一章将全书涉及的重点科学词汇进行了梳理和总结，方便孩子理解和记忆科学词汇。

2024年，《培生科学虫双语百科》系列双语版由我社首次引进出版。为了帮助青少年读者进行高效的独立阅读，并方便家长进行阅读指导或亲子共读，我们为本书设置了以下内容。

（1）每个分册第一部分的英语原文（奇数页）后均配有对应的译文（偶数页），跨页部分除外。读者既可以进行汉英对照阅读，也可以进行单语种独立阅读。问题前面的 📖 符号表示该问题可在第二部分预留的位置作答。

（2）每个分册第二部分的电子版译文可在目录页扫码获取。

（3）本书还配有英音朗读音频和科学活动双语视频，也可在目录页扫码获取。

最后，祝愿每位读者都能够享受双语阅读，在汲取科学知识的同时，看见更大的世界，成为更好的自己！

电子工业出版社青少年教育分社
2024年1月

Contents 目录

Part 1　科学虫趣味课堂　　　　　　　　　　　／1

Part 2　科学虫大闯关　　　　　　　　　　　　／33

Part 3　科学词汇加油站　　　　　　　　　　　／49

Part2译文

配套音视频

Electricity

Electricity is really useful. It can make machines move, heat things up, make things light up or make sound.

Word Box
electricity
lightning

This lightning is electricity moving between a cloud and the ground or a cloud and something on the ground such as a building or tree.

1. Discuss with your group what is happening to the tree in this picture.
2. Why can it be dangerous to go out in a lightning storm?

Electricity is a natural thing in the world around us. The Ancient Greeks knew about electricity over 2,000 years ago! It is only in the last 100 years or so that we have been able to safely generate electricity that is useful to us.

This picture shows a power station generating electricity.

Discuss with your group what you know about electricity. Think of all the things you use that use electricity.

1. How does having electricity help make our lives easier? Record what is good about having electricity in our homes. Are there any disadvantages? Record these too.
2. What four things would you miss if we could not use electricity? Explain why you would miss them.

电

电是十分有用的。它能让机器运转起来,能加热物体,能带来光亮,也能发出声音。

单词表
electricity 电;电流
lightning 闪电

闪电是云和地面之间或是云和地面上的物体(如建筑物或树)之间的放电现象。

1. 和你的小伙伴一起,讨论一下图中的树发生了什么。
2. 在雷电天气外出为什么会很危险?

电是存在于我们周围的一种自然物质。在2000多年前,古希腊人就已经了解电这种物质了!但直到最近100年左右,我们才能安全地生产为人类所用的电。

这张照片显示了正在发电的发电站。

和你的小伙伴一起,讨论一下,你了解的关于电的知识。思考一下有哪些你使用过的需要用电的物品。

1. 电的存在如何让我们的生活变得更加轻松?记录一下家里有电为我们带来的好处。电有哪些缺点呢?也记录下来吧。
2. 如果在我们的生活中没有电,你会想念哪四样物品?解释一下原因吧。

Electrical Appliances

An electrical **appliance** is an object that uses electricity to work. Some appliances use **mains electricity** so need to be plugged into a socket in the wall. The socket has holes that allow the plug on the electrical appliances to be connected into mains electricity.

Word Box
appliance
battery
mains electricity

Some appliances use electricity from a **battery**. Some appliances can use both. Mains electricity and batteries are called an electrical supply.

Sorting electrical appliances

You will need: sticky notes and catalogues or magazines to cut up.

1. Try to identify an electrical appliance for every letter of the alphabet, and record each one on a separate sticky note. You have 10 minutes to get as many letters as possible completed.
2. Compare with the other groups in the class to see which group has found appliances for the most letters.
3. Use catalogues, household magazines, and the internet if available, to try to find an appliance for any letters that no group has managed to find one for.
4. Sort images of appliances from catalogues and magazines into those that use mains electricity, those that use batteries and those that can use either.

电器

电器是指需要用电工作的**器具**。一些电器需要接上**干线供应的电力**，所以要插进墙上的插座。插座上的洞能让电器上的插头连上干线供应的电力。

单词表
appliance 器具
battery 蓄电池
mains electricity 干线供应的电力

一些电器用的电来源于**蓄电池**。而另外一些电器则既能接干线供应的电力又能用蓄电池供电。干线供应的电力和蓄电池都被称为电源。

将电器分类

你将需要：便利贴、用来裁剪的手册或杂志。

1 试着为字母表中的所有字母都想出一个对应的电器，各用一张便利贴记录下来。你有10分钟的时间来完成，想出的电器的种类越多越好。

2 和其他的小伙伴一起，比较一下，看看谁想出的电器种类最多。

3 如果有的字母没有人想出对应的答案，试着利用可用的产品目录、家庭杂志或者互联网查找一下。

4 将产品目录和杂志中的电器图片分类，分为使用干线供应的电力、使用蓄电池或使用两者都可以三类。

We say something is manual if it is done by hand. Manual objects do not use electricity to work.

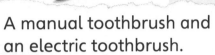

A manual toothbrush and an electric toothbrush.

1 In your group discuss whether you would prefer to use a manual toothbrush or an electric toothbrush. Explain why.

Working as a pair you have 10 minutes to list as many appliances as you can that can be both electrical and manual (does not use electricity).

Share your ideas to help build a class list.

These pictures show some electrical appliances that are often used at home.

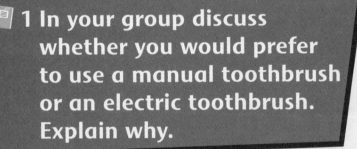

2 How many do you use in your home?

3 Discuss with your group what you would do if they had never been invented.

4 Record all the electrical appliances you use in a day. For each one describe what you would do if the appliance did not exist.

如果某件事可以通过手工完成，我们称它是手动的。手工物品不需要用电就可以工作。

一只手动牙刷和一只电动牙刷。

1 和你的小伙伴一起，讨论一下，你更愿意使用手动牙刷还是电动牙刷。解释一下原因吧。

你有10分钟的时间和你的小伙伴一起，列出既能用电工作又能手动（不需要用电工作）的器具。

分享你们的想法，列一张汇总的表格。

这些图片展示了人们在家经常用到的电器。

2 哪些电器是你家在用的？

3 和你的小伙伴一起，讨论一下，如果这些电器没有被发明出来，你会怎么办。

4 记录你在一天中用过的所有电器。分别描述一下，如果这些电器不存在，你会怎么办。

Electrical Inventions

Electricity isn't just used to make lights work and kettles boil. Many wonderful and unusual **inventions** use it too.

Word Box
invention

The electrically heated jacket

In 1932, the traffic police in America had a problem. They had to stand outside for a long time in the cold weather. So they made an electrically heated vest!

1 Discuss whether you think it is dangerous to wear a jacket connected to an electricity supply.

There were points in the street where the police officers could connect their vests to an electricity supply and warm themselves up.

The Baker electric car

In 1909 Walter Baker invented an electric car. The car was powered by batteries and could travel 160 kilometres at up to 40 km per hour before it needed recharging.

2 Discuss any advantages or disadvantages of an electric car with your group.

Over 100 years later, electric cars are now common all over the world.

与电相关的发明

电不仅仅是用于电灯和电热水壶的。很多奇妙而不同寻常的**发明**也会利用电。

单词表
invention 发明

电热夹克

1932年，美国的交警遇到了一个难题。他们要在寒冷的天气里在室外停留很长一段时间，所以他们制作了电热马甲！

街上有一些站点可以让警员们将马甲连接到电源上，让他们的身体热起来。

1 讨论一下，你认为穿上一件连接电源的夹克有没有危险。

贝克电动汽车

1909年，沃尔特·贝克（Walter Baker）发明了电动汽车。这辆车由电池供电，充满一次电后，能以40千米每小时的速度行驶160千米。

100多年后，电动汽车在全世界变得常见。

2 和你的小伙伴一起，讨论一下电动汽车的优点和缺点。

The smart wheel for bikes

This recent invention makes cycling much easier! The rear wheel of a bicycle is replaced by an electric wheel. This is good news for people who have to cycle up hills.

The electric wheel can travel at 40 kmph for about 80 km without recharging!

Design a new electrical appliance that would make life easier for you.

1 Draw and label your new appliance.

2 Write a sentence to describe what it does.

3 Discuss with your group the parts of this toaster you think might still be used on modern toasters.

100 years ago toast would not pop up when it was ready. The first automatic toaster was not invented until 1919.

Some people think that electricity is the greatest discovery of all time.

1 Why do you think they think that? Do you agree? Discuss your ideas with your group.

2 How many reasons can you think of? Make a list.

3 Think of another great discovery.

自行车的智能轮胎

这项最新的发明让骑自行车变得更加轻松！这辆自行车的后轮是由一个电动轮胎代替的。对于那些不得不骑上坡的人来说，这真是个好消息。

电动轮胎充一次电就能以40千米每小时的速度行驶80千米！

设计一个能让你的生活变得更加轻松的新电器。

1. 画出并标注你的新电器。
2. 用一句话描述它的作用。

3 和你的小伙伴一起，讨论一下，图片中烤面包机的哪些部分还保留在现代烤面包机中。

100年前，面包烤好后并不会自动地弹起来。直到1919年，第一个自动烤面包机才被发明。

一些人认为电是有史以来最伟大的发明。

1. 你认为人们这样想的原因是什么？你同意这个观点吗？和你的小伙伴讨论一下。
2. 你能想出多少种原因？请列出来。
3. 尝试想出其他的伟大发明。

Electrical Components

A **circuit** is a path that electricity takes. An electrical appliance needs a complete circuit for it to work. It needs a source of electricity such as a **cell** or battery (two or more cells joined together).

A **component** is a part of a circuit such as a **bulb**, **buzzer** or **wires**. There are many different types of component. Each one has a different use in a circuit.

Word Box
bulb
buzzer
cell
circuit
component
switch
wire

These pictures show some electrical components.

1 Identify the electrical components in the pictures with your partner.

Lighting a bulb

You will need: bulb, bulb holders, wires, cell, sticky notes, modelling material...

Work with a partner. Light the bulb.

2 In your group discuss when you use switches.
3 Do all electrical appliances have switches?
4 Why are switches useful in circuits?

电路元件

电路是电流经过的路线。电器需要有完整的电路才能正常工作。电器需要电源，例如，单个**电池**或蓄电池（由两个或多个电池串联组成）。

元件是电路的某个部分，例如，**灯泡**、**蜂鸣器**或**导线**。有很多不同类型的电路元件，每种元件在电路中都起着不同的作用。

> **单词表**
> bulb 灯泡
> buzzer 蜂鸣器
> cell 电池
> circuit 电路
> component 元件
> switch 开关；转变
> wire 导线

这些图片展示了一些电路元件。

📖 **1** 和你的小伙伴一起，指出图中的电路元件分别是什么。

点亮灯泡

你将需要：灯泡、灯座、导线、电池、便利贴、模型材料……

和你的小伙伴一起，试着点亮灯泡。

2 和你的小伙伴一起，讨论一下，在什么时候你会使用开关。
3 所有的电器都有开关吗？
4 为什么开关在电路中有用？

Making a buzzer buzz

You will need: bulb, buzzer, wires, cell, sticky notes, sticky tape, scissors, card and recycled materials for model making, including cardboard tubes, boxes...

1 Work with a partner. Make the buzzer buzz.
2 What happens when you **switch** over the wires attached to the buzzer?
3 Use the recycled materials, such as cardboard tubes and boxes, to make a model that shows how you have built a circuit and made the buzzer buzz. Label the electrical components in the model with sticky notes.

Electricity is only useful to us when we can control it. One way of controlling electricity is by putting a switch into a circuit.

These pictures show different types of switches used in circuits.

Lighting a bulb that switches on and off

You will need: bulb, bulb holders, wires, cell, a switch...

1 Work with a partner. Construct a circuit that allows a bulb to be switched on and off.
2 Draw a labelled diagram showing this circuit. Then draw a labelled diagram showing how you think you could turn a buzzer on and off using a switch.

让蜂鸣器鸣叫

你将需要：灯泡、蜂鸣器、导线、电池、便利贴、胶带、剪刀、卡片、用来制作模型的可回收材料，包括硬纸管、箱子……

1. 和你的小伙伴一起，让蜂鸣器鸣叫。
2. 当你把导线转接到蜂鸣器时，会发生什么？
3. 使用可回收材料，例如，硬纸管和箱子来制作一个模型，展示一下你是如何搭建电路的，以及你是如何让蜂鸣器鸣叫的。用便利贴标注模型中的电路元件。

电只有在为我们所控制的情况下，才对我们有用。控制电的一个方法就是在电路中设置开关。

这些图片展示了电路中用到的不同开关。

用开关控制灯泡

你将需要：灯泡、灯座、导线、电池、一个开关……

1. 和你的小伙伴一起，搭建一个能用开关控制灯泡的电路。
2. 画出并标注电路示意图。然后再画出一个示意图进行标注，展示一下你认为如何用开关打开或关闭蜂鸣器。

Completing Circuits

1. Which circuit will have a bulb that is lit?
2. Why would the bulb in the other circuit not light up?

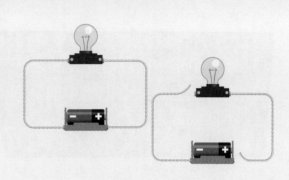

Complete circuits

You will need: wires with crocodile clips, bulbs, bulb holders, cells...

1. Work in a small group. Draw a circuit for a partner to build. Ask them to predict whether it will work or not, and to say why.
2. Now build the circuit. Was their prediction correct?
3. Discuss with your group what you have discovered.
4. Come up with a simple rule to use to predict whether a bulb will light or not.

Light a bulb

Work in a pair to make a bulb light up using the smallest possible number of wires.

📓 Record how you did this.

完成电路

1 哪个电路中的灯泡能亮起来?
2 为什么另外一个电路中的灯泡不能亮起来?

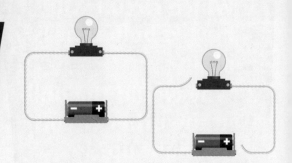

完成电路

你将需要:带鳄鱼夹的导线、灯泡、灯座、电池……

1 和你的小伙伴一起,进行实验。为了让你的小伙伴搭建电路,你要先画一个电路。让你的小伙伴预测电路是否能正常工作,并解释原因。
2 现在搭建电路吧。小伙伴的预测是正确的吗?
3 和你的小伙伴一起,讨论一下你们的发现。
4 思考一下有没有一条简单的规则可以用来预测灯泡亮不亮。

点亮灯泡

和你的小伙伴一起,用尽可能少的导线让灯泡亮起来。

📖 记录一下你们是如何做的。

Make Electricity Spider Puzzles

We know that electricity needs an unbroken path or circuit to pass through to make bulbs light up, buzzers buzz and appliances work. Use this knowledge to make a game.

Making an electricity spider

We can make an eight-legged spider with whole wires or cut wires for legs.

You will need: wires of four different colours, modelling dough, wire cutters, components of your choice to make a circuit to test whether the legs of the spiders are carrying electricity.

1 Take four 20 cm lengths of different coloured wire.
2 Cut one length of wire in half in the middle.
3 Place all the wires into a blob of modelling dough (the spider's body) so that you have a spider with eight legs (two of each colour).

Now construct a circuit into which you can put the spider to find out which legs are cut in the middle.

Electricity spider puzzle

1 With a partner make another puzzle spider. This time cut *two* different coloured wires.
2 Challenge another pair of classmates to spot which legs are cut in the middle.

制作电蜘蛛谜团

我们知道电流需要经过完整的回路才能让灯泡亮起、让蜂鸣器鸣叫、让器具正常运行。运用这条规则设计一个游戏。

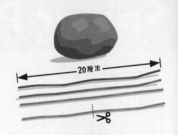

制作一只电蜘蛛

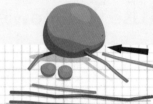

我们可以用完整的导线或裁切好的导线当作腿制作一只八条腿的蜘蛛。

你将需要:四种不同颜色的导线、模型面团、剪线钳、你自行选择的电路元件,这些电路元件搭建的电路用来测试蜘蛛的腿是否通电。

1. 准备四根不同颜色的20厘米长的导线。
2. 将其中一根导线对折并从中间剪断。
3. 将所有导线插入模型面团中(蜘蛛的身体),这样我们得到一只有八条腿的蜘蛛(每种颜色的导线对应两条腿)。

现在搭建一个电路,你可以把蜘蛛放进检测电路中,找出蜘蛛的哪两条腿是从中间剪断的。

电蜘蛛谜团

1. 和你的小伙伴一起,再做一只蜘蛛,这次剪断两条不同颜色的导线。
2. 和另外一对小伙伴比赛,找出哪几条腿是从中间剪断的。

Electrical Conductors and Insulators

An electrical **conductor** allows electricity to pass through it. An electrical **insulator** does not allow electricity to pass through it.

Word Box
conductor
insulator

investigating electrical conductors and insulators

You will need: batteries, bulbs, bulb holders, wires with crocodile clips attached, and spoons made out of metal, wood, plastic...

1 In a small group construct a circuit that lights up a bulb. You are going to test each spoon to find out whether it conducts electricity. Record your results.

2 Discuss with your group, what another group's results show.

Look closely at the kitchen utensils in this picture.

1 Decide which will act as electrical insulators and not conduct electricity. Draw these and say why you have chosen them.

2 What do conductors of electricity all have in common?

导体与绝缘体

导体可以让电流通过。
绝缘体不允许电流通过。

单词表
conductor 导体
insulator 绝缘体

探究导体与绝缘体

你将需要：电池、灯泡、灯座、附有鳄鱼夹的导线、金属勺子、木头勺子、塑料勺子……

1 和你的小伙伴一起，搭建一个让灯泡亮起来的电路。测试不同材质的勺子能否导电。记录你的结果。

2 和你的小伙伴一起，讨论一下实验结果。

仔细观察右图中的厨房用具。
1 选出绝缘体，把这些绝缘体画出来并解释你这样选择的原因。
2 导体有哪些共同点？

conducting objects

Collect 15 small objects made from a variety of different materials. Include as many of the ones in the picture as you can.

Work in a small group to identify what material each of the objects is made from.

📖 1 Predict whether each object is made from a material that conducts electricity or a material that is an electrical insulator.

2 Make a circuit similar to the one you used to test the spoons in the previous activity.

📖 3 Test your objects to check your predictions.

1 Identify anything on the clothes in the picture that would conduct electricity.

2 As a group discuss what you would and wouldn't wear if you had to go out in a lightning storm. Why?

导体

收集由不同材料制成的15件小物品。尽可能多地涵盖图中所示的材料。

和你的小伙伴一起，识别每件物品是由哪些原材料组成的。

1. 预测一下组成每件物品的原材料是导体还是绝缘体。
2. 搭建一个在上一节用到的检测勺子是否导电的相似电路。
3. 将你的物品一一进行测试，验证一下预测是否正确。

---

1. 在图中所示的衣服上，请指出哪些东西可以导电。
2. 和你的小伙伴一起，讨论一下，如果你在雷电天外出，你会穿什么？不会穿什么？为什么？

Science Skills

Investigating Electrical Conductors – Investigate it!

To learn more about something, we explore and investigate, and this is an important part of science. For example, here we investigate how different materials behave with water.

Word Box
method

Material and water

You will need: samples of materials similar to those in the pictures above, a cup of water...

1 In a small group explore the samples of materials. Discuss how they are similar and different.

2 Explore what happens when you dip a sample of each material in water.

There are several investigations you could carry out to discover whether the materials in the pictures conduct electricity or are electrical insulators. Investigations start with a question. For example, are all types of paper electrical insulators?

1 List as many questions as you can to discover whether materials conduct electricity or are electrical insulators.
2 Record them on sticky notes.
3 Share some of your favourite questions.

科学技能

探究导体——探究一下吧！

为了更多地了解某个事物，我们需要进行探索与探究，这正是科学研究的重要部分。例如，我们现在就在探究不同材料是如何与水发生作用的。

单词表
method 方法

材料与水

你将需要：与上面图片上的物体类似的材料样本、一杯水……

1. 和你的小伙伴一起，探索这些材料样本。讨论一下它们的相似之处和不同之处。
2. 当你将某个材料样本浸入水中时，探索一下会发生什么。

你可以进行不同的探究来发现图中的材料是导体还是绝缘体。探究都需要先提出一个问题，例如，所有纸质材料都是绝缘体吗？

1. 请列出尽可能多的问题，用这些问题来判断材料是导体还是绝缘体。
2. 在便利贴上记录下来。
3. 分享你最感兴趣的一些问题。

investigating electrical conductivity

You can now answer one of the questions by carrying out an investigation.

1. Stick your question at the top of a large sheet of paper then exchange your paper with another group.
2. Record instructions on how you can find the answer to their question by carrying out an investigation.
3. Retrieve your own question. Read through the investigation instructions. Discuss any changes you think need to be made. You can make notes about these on the paper in a different colour. This is the way you do your investigation, called the **method** of the experiment.
4. Record the question you are going to investigate and the method. You can use drawings and diagrams as well.

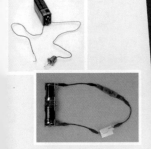

These two pictures show experiments looking at whether different types of papers conduct electricity.
Do you think this is a good way to find out? Why?

5. Discuss what equipment you will need to complete your investigation and make a list.
6. Record what you predict will happen.
7. Carry out the investigation. Record your results.
8. Discuss what you found out. These are your conclusions.

探究导电性

你可以通过实验探究来回答其中的问题。

1. 将你的问题粘在一大张纸的顶部,然后同其他的小伙伴交换纸张。
2. 写下一些提示,提醒你自己要如何通过探究来找到问题的答案。
3. 将纸张交换回来,通读实验指导。讨论需不需要做一些改动。你们可以用不同颜色的笔在纸上做笔记。这就是你们进行探究的方式,叫作实验方法。
4. 记录你们要探究的问题和方法。可以借助图画和图表。

这两张图片展示了测试不同类型的纸是否导电的实验。
你认为这是探究的好方法吗?为什么?

5. 讨论一下你们需要哪些设备来进行探究,并列出表格。
6. 记录你们的预测。
7. 进行实验。记录实验结果。
8. 讨论你们的探究结果。这些结果就是你们得到的结论。

Science Skills

Using Switches – Make it!

1 What do these pictures show?
2 What would these switches do?

Using switches

You will need: different types of switches similar to the ones in the picture, bulbs, bulb holders, buzzers, batteries, wires with crocodile clips...

Work in a group to observe what happens when you use switches in simple circuits.

1 Make a circuit that lights a bulb and one that makes a buzzer buzz.
2 Explore adding switches in these circuits. Discuss what happens with each one.

We can make our own switches.

Step 1

Step 2

Step 3

科学技能

使用开关——
制作一下吧!

1 这些图片展示了什么?
2 这些开关有什么作用?

使用开关

你将需要:与图中类似的不同类型的开关、灯泡、灯座、蜂鸣器、电池、带鳄鱼夹的导线……

和你的小伙伴一起,观察在简单电路中使用开关会发生什么。

1 制作一个能点亮灯泡的电路和能让蜂鸣器鸣叫的电路。
2 探究在电路中添加不同的开关。讨论一下之后每个电路会发生什么。

我们能自己动手制作开关。

步骤1

步骤2

步骤3

Science Skills

Make it!

Follow these instructions to make a switch.

To make a switch you will need:

- two split pins
- a metal paperclip
- a piece of card that measures 8cm long and 4cm wide
- a ruler

1. Measure the paperclip and make two small holes near the middle of the card that are the length of the paperclip apart.

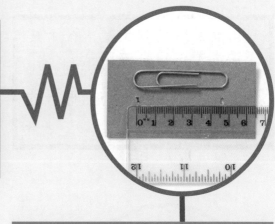

2. Push the split pins through the holes and splay the prongs on the reverse of the card to hold the pins in place.

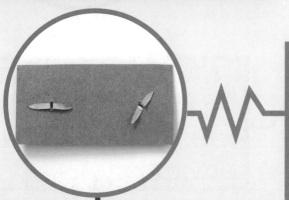

3. Hook one end of the paperclip over the top of one of the split pins. Check that the other end of the paperclip can be moved to touch the other split pin.

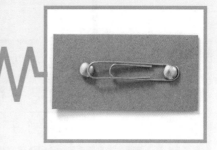

科学技能

制作一下吧!

按照下面的说明制作一个开关。

为了制作一个开关,你将需要:

- 两个开口销
- 一个金属回形针
- 一张长8厘米,宽4厘米的卡片
- 一把直尺

1.测量回形针的长度,在卡片宽度的一半处戳两个小洞,小洞之间的距离刚好与回形针的长度相等。

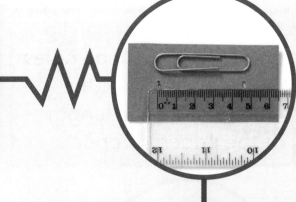

2.将两个开口销分别插入小洞中并使其张开,在卡片的背面打开开口销,将开口销固定在合适的位置。

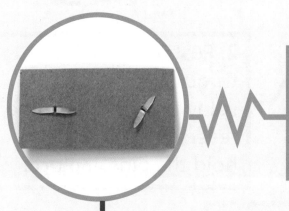

3.用回形针的一端勾住开口销的顶部,观察回形针的另一端是否可以移动并能贴上另一个开口销。

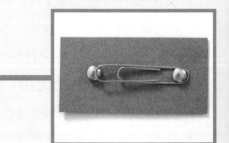

Science Skills

Test your switch to check it works using these instructions.

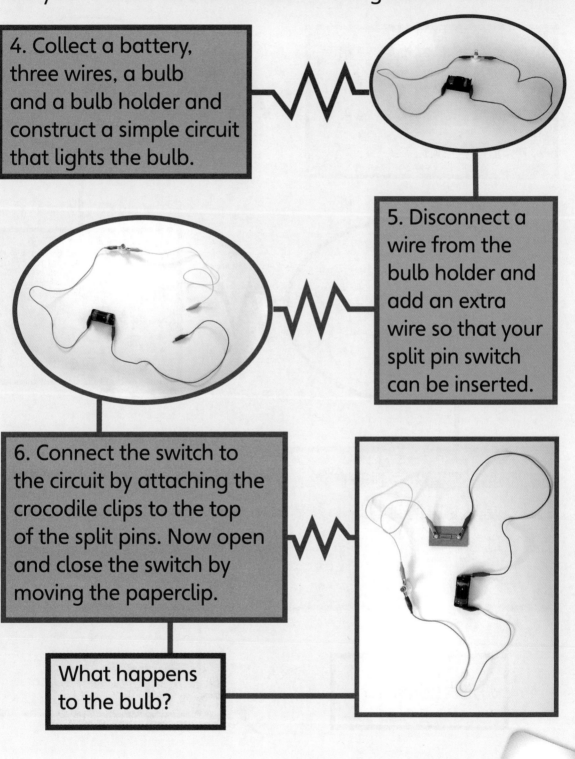

4. Collect a battery, three wires, a bulb and a bulb holder and construct a simple circuit that lights the bulb.

5. Disconnect a wire from the bulb holder and add an extra wire so that your split pin switch can be inserted.

6. Connect the switch to the circuit by attaching the crocodile clips to the top of the split pins. Now open and close the switch by moving the paperclip.

What happens to the bulb?

科 学 技 能

按照下面的说明测试你的开关是否能正常工作。

4.收集一节蓄电池、三根导线、一个灯泡和一个灯座,用两根导线组成一个能点亮灯泡的简单电路。

5.将一根导线与灯座断开,再加上另外一根导线,这样可以放入开口销开关。

6.通过将鳄鱼夹放到开口销顶端,使开关连接到电路中。然后通过移动回形针来闭合或断开开关。

灯泡发生了什么?

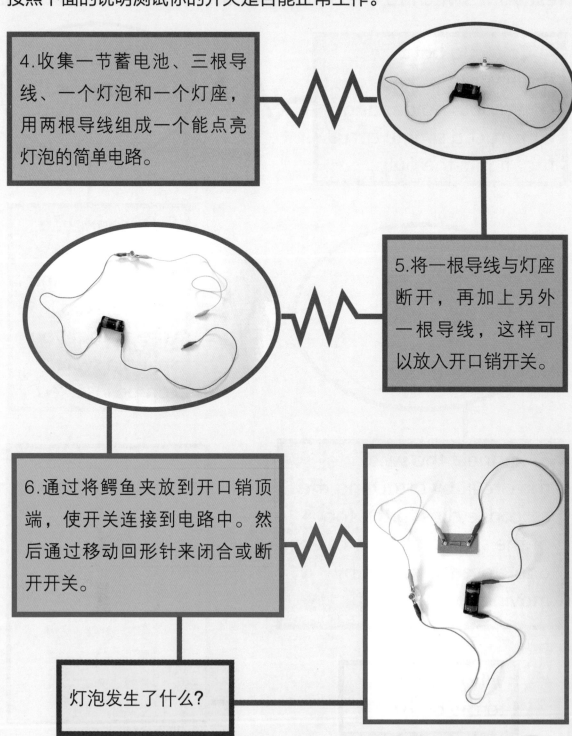

Electricity

1 Record in this table some advantages and disadvantages of having electricity.

Advantages of electricity	Disadvantages of electricity

2 Draw and label four appliances that you would miss if we could not use electricity. Explain why you would miss them.

I would miss _____ because	I would miss _____ because	I would miss _____ because	I would miss _____ because

Electrical Appliances

1. Gather together images of appliances from catalogues and magazines. Sort them into appliances that use batteries, ones that use mains electricity and those that can use either. Record the way you sorted these.

Mains, battery or both
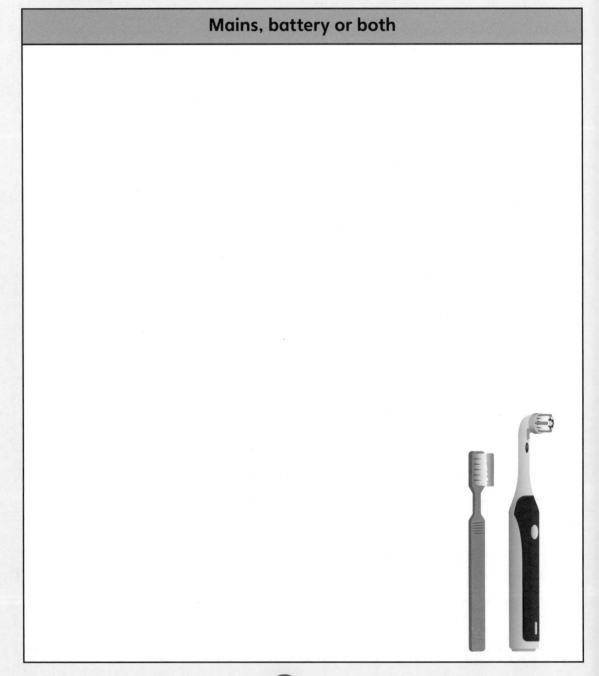

2 There are both manual and electrical toothbrushes. List three reasons people might choose to buy and use a manual toothbrush rather than an electrical one.

_____

_____

_____

3 Record all the electrical appliances you use in a day. For each one describe what you would do if the appliance did not exist.

Electrical appliance	If it did not exist I would ...

Electrical Inventions

Design a new electrical appliance that would make life easier for you.

1 **Draw and label your invention and give it a name.**

2 **Write a sentence describing what it does.**

3 **Explain why it would be a great invention.**

Some people think that electricity is the greatest discovery of all time.

4   Why do you think they believe this? Record as many reasons as you can.

---

5   Look at these everyday objects – all of these were invented at some point in the past. Tick all the ones that use electricity.

6   Add three more objects that use electricity.

Electrical Components

1. Write the name of these electrical components.

   _____

2. Match the component to its description by drawing a line between them.

bulb	An electrical component that makes a noise (buzzes) when electricity passes through it.
battery (cell)	An electrical component that lights up when electricity passes through it.
buzzer	Two or more cells placed end to end in a circuit. Produces electricity.

3. If you have real circuit components available observe them closely, ideally with a hand lens. Draw the inside of a bulb.

4 Draw a labelled diagram of a circuit that allows a bulb to be switched on and off. You may have constructed one yourself that you can draw here.

5 Draw a labelled diagram showing how you think you could turn a buzzer on and off using a switch.

Completing Circuits

1. Draw four circuits. Draw three circuits with mistakes in them. These circuits will not make the buzzer sound. Draw one that will work. Ask a classmate to work out which one will work.

Circuit A	Circuit B
Circuit C	Circuit D

2. Draw a picture to show how you made a bulb light up using the smallest possible number of wires.

3. Make an eight-legged spider with modelling dough and whole wires or cut wires for legs. Then test it to find which wires are cut. Draw and write a set of instructions for making a spider puzzle with two wires cut.

Diagram	Instruction
	_____ _____ _____
	_____ _____ _____
	_____ _____ _____
	_____ _____ _____

Electrical Conductors and Insulators

We can test a range of materials to see whether they conduct electricity, and so discover whether the material is a conductor or an insulator. You may have carried out experiments with metal, wood and plastic, or you can research these materials. Then answer these questions.

1. **What do you think would be similar about all objects that conduct electricity?**

   _____
   _____

2. **What do you think would be similar about all objects that are electrical insulators?**

   _____
   _____

3. **Name an object that could be used as a switch in an electrical circuit. Why could you use this object?**

   _____
   _____

4. **Draw and label some kitchen utensils that you think will act as electrical insulators, and not conduct electricity.**

Which Objects Conduct Electricity?

Predict whether each object you collected on page 21 part 1 is made from a material that conducts electricity or a material that is an electrical insulator.

Record your predictions in this table. If you have the equipment to build a circuit including the object, you could also add the result to the table, or research each material.

Object	Material(s) it is made from	Prediction: Conducts electricity? Yes/No	Result: Conducts electricity? Yes/No

Investigating Materials

1. List three questions about materials and electrical conductivity that you could investigate.

2. Choose one question to investigate.

   Method (describe what you are going to do in words and pictures)

3. Use these diagrams to help list the equipment you will use to answer your question.

4. Now record any predictions you made.

I think that ...	because ...

5. Write down your conclusions. I found that ...

Making a Switch

You will need: a selection of recycled materials including wire, aluminium foil, paper clips, thin card, also wire strippers, spring-loaded clothes pegs, sponges and insulating tape.

1. **Work as a pair to design a switch and record the design. Label your diagram.**

   Our design for a switch

2. **Explain why you think it will work.**

Testing the Switch

Give instructions on how to test the switch you have designed. Use pictures, labels and text.

How to test the switch

The Wonders of Electricity

You are going to explain how wonderful electricity is to a six-year-old child. You need to tell him:

- What electricity does
- Why it can be dangerous
- How it can be used to make things work
- What the world would be like without it

Write down what you would say to him.

Glossary

appliance (electrical): object that uses electricity to work

battery: produces electricity

bulb: electrical component that lights up when electricity passes through it

buzzer: electrical component that makes a noise (buzzes) when electricity passes through it

cell: produces electricity. Two or more cells placed end to end in a circuit is called a battery

circuit: a cell, battery, wires and other electrical components all joined in a loop that electricity can flow around

component (electrical): part of a circuit such as a bulb, buzzer or wire

conductor (electrical): a material that allows electricity to flow through it

electricity: a form of energy that occurs naturally or is produced and passed through wires to make electrical appliances work

insulator (electrical): a material that does not allow electricity to flow through it

invention: a new device (object)

lightning: natural electricity from a cloud towards the ground

mains electricity: electricity supplied to schools and houses

method: how something is done, usually in steps

switch: an electrical component that causes a break in a circuit

wire: allows electricity to pass when used in a complete circuit

词汇表

器具（与电有关的）：需要用电才能正常工作的物体

蓄电池：能供电

灯泡：电流通过时能被点亮的电路元件

蜂鸣器：电流通过时能发出噪声（蜂鸣声）的电路元件

电池：能产生电。由两节或两节以上的电池串联组成的电路叫作蓄电池

电路：电流能通过的闭合回路，包括电池、蓄电池、导线等电路元件

元件（与电有关的）：电路中的一部分，例如，灯泡、蜂鸣器和导线

导体（与电有关的）：能让电流通过的材料

电：自然界中产生的或人为制造的一种能量形式，能通过导线进行流动，可以让电器正常运行

绝缘体（与电有关的）：不能让电流通过的材料

发明：一种新出现的设备（物体）

闪电：由云层射向地面的自然界中的电

干线供应的电力：供应给学校和家庭的电

方法：某件事是如何完成的，通常是分步骤进行的

开关：能够让电路闭合或断开的电路元件

导线：在闭合回路中能让电流通过

培生科学虫双语百科
奇妙物理

Sound

声音

英国培生教育出版集团 著·绘
徐昂 译

电子工业出版社
Publishing House of Electronics Industry
北京·BEIJING

Original edition, entitled SCIENCE BUG and the title Sound Topic Book, by Debbie Eccles published by Pearson Education Limited © Pearson Education Limited 2018
ISBN: 9780435197018

All rights reserved. No part of this book may be reproduced or transmitted in any form or by any means, electronic or mechanical, including photocopying, recording or by any information storage retrieval system, without permission from Pearson Education Limited.

This adaptation of SCIENCE BUG is published by arrangement with Pearson Education Limited.
Chinese Simplified Characters and English language (Bi-lingual form) edition published by PUBLISHING HOUSE OF ELECTRONICS INDUSTRY, Copyright © 2023.
For sale and distribution in the mainland of China exclusively (except Hong Kong SAR, Macau SAR and Taiwan).

本书中英双语版由Pearson Education（培生教育出版集团）授权电子工业出版社在中华人民共和国境内（不包括香港、澳门特别行政区及台湾地区）独家出版发行。未经出版者书面许可，不得以任何方式抄袭、复制或节录本书中的任何部分。

本套书封底贴有 Pearson Education（培生教育出版集团）激光防伪标签，无标签者不得销售。

版权贸易合同登记号　图字：01-2022-2381

图书在版编目（CIP）数据

培生科学虫双语百科. 奇妙物理. 声音：英汉对照 / 英国培生教育出版集团著、绘；徐昂译. --北京：电子工业出版社，2024.1
ISBN 978-7-121-45132-4

Ⅰ.①培… Ⅱ.①英… ②徐… Ⅲ.①科学知识-少儿读物-英、汉 ②声-少儿读物-英、汉 Ⅳ.①Z228.1 ②O42-49

中国国家版本馆CIP数据核字（2023）第035078号

责任编辑：李黎明　文字编辑：王佳宇
印　　刷：河北迅捷佳彩印刷有限公司
装　　订：河北迅捷佳彩印刷有限公司
出版发行：电子工业出版社
　　　　　北京市海淀区万寿路173信箱　邮编：100036
开　　本：787×1092　1/16　印张：35　字数：840千字
版　　次：2024年1月第1版
印　　次：2024年2月第2次印刷
定　　价：199.00元（全9册）

凡所购买电子工业出版社图书有缺损问题，请向购买书店调换。若书店售缺，请与本社发行部联系，联系及邮购电话：（010）88254888，88258888。
质量投诉请发邮件至zlts@phei.com.cn，盗版侵权举报请发邮件至dbqq@phei.com.cn。
本书咨询联系方式：010-88254417，lilm@phei.com.cn。

使用说明

欢迎来到少年智双语馆！《培生科学虫双语百科》是一套知识全面、妙趣横生的儿童科普丛书，由英国培生教育出版集团组织英国中小学科学教师和教研专家团队编写，根据英国国家课程标准精心设计，可准确对标国内义务教育科学课程标准（2022年版）。丛书涉及物理、化学、生物、地理等学科，主要面向小学1~6年级，能够点燃孩子对科学知识和大千世界的好奇心，激发孩子丰富的想象力。

本书主要内容是小学阶段孩子需要掌握的物理知识，含9个分册，每个分册围绕一个主题进行讲解和练习。每个分册分为三章。第一章是"科学虫趣味课堂"，这一章将为孩子介绍科学知识，培养科学技能，不仅包含单词表、问题和反思模块，还收录了多种有趣、易操作的科学实验和动手活动，有利于培养孩子的科学思维。第二章是"科学虫大闯关"，这一章是根据第一章的知识点设置的学习任务和拓展练习，能够帮助孩子及时巩固知识点，准确评估自己对知识的掌握程度。第三章是"科学词汇加油站"，这一章将全书涉及的重点科学词汇进行了梳理和总结，方便孩子理解和记忆科学词汇。

2024年，《培生科学虫双语百科》系列双语版由我社首次引进出版。为了帮助青少年读者进行高效的独立阅读，并方便家长进行阅读指导或亲子共读，我们为本书设置了以下内容。

（1）每个分册第一部分的英语原文（奇数页）后均配有对应的译文（偶数页），跨页部分除外。读者既可以进行汉英对照阅读，也可以进行单语种独立阅读。问题前面的 📖 符号表示该问题可在第二部分预留的位置作答。

（2）每个分册第二部分的电子版译文可在目录页扫码获取。

（3）本书还配有英音朗读音频和科学活动双语视频，也可在目录页扫码获取。

最后，祝愿每位读者都能够享受双语阅读，在汲取科学知识的同时，看见更大的世界，成为更好的自己！

电子工业出版社青少年教育分社
2024年1月

Contents 目录

Part 1　科学虫趣味课堂　　　　　　　　　　　　　　／1

Part 2　科学虫大闯关　　　　　　　　　　　　　　　／35

Part 3　科学词汇加油站　　　　　　　　　　　　　　／51

Part2译文

配套音视频

What Do We Know about Sound?

Close your eyes and sit quietly for a few moments. You will hear many different sounds. The sounds that enter our ears are all made by something. That 'something' is the sound **source**.

Word Box
source

📖 1a List the sounds you can hear.
📖 1b What made these sounds?

Look at the pictures.
2 Which of these sounds do you like best? Why?
3 Which of these sounds do you like least? Why?

The sound source is not usually right next to your ears but you can still hear the sound it makes.

📖 4 How do you think you can hear sounds from a sound source that is not next to you?

关于声音，我们知道什么？

闭上双眼安静地坐一会儿。你会听到许多声音。这些进入我们耳朵的声音都是由某种物体发出的。发出声音的物体叫作声源。

单词表
source 来源

📖 1a 列举你能听到的声音。
📖 1b 这些声音是由什么发出的？

观察上面的图片。
2 你最喜欢哪种声音？为什么？
3 你最不喜欢哪种声音？为什么？

声源通常离你有一定的距离，但是你仍然能听到声源发出的声音。

📖 4 你认为，怎样才能听见远处的声源发出的声音呢？

Science Skills

Classify it!

There are many different types of sounds. We find some much nicer to listen to than others. We sometimes call unpleasant sounds **noise**.

Word Box
high
loud
low
noise
quiet
volume

1 Talk to a partner. What types of sound would you describe as noise?

A fire alarm

A dog barking

A flute being played

Sounds can be different **volumes**. The volume is how **loud** or **quiet** a sound is. Some sounds can easily be heard. These are loud sounds. Others are much more difficult to hear because they are quiet.

Guitar music

2a When do you use a loud voice?
2b When do you use a quiet voice?

A whistle

科学技能

分类一下吧！

声音有很多种类。我们发现有些声音比其他声音更好听。我们通常把不太悦耳的声音称为**噪声**。

单词表
high 高的
loud 大声的
low 低的
noise 噪声
quiet 小声的
volume 响度

1 和你的小伙伴一起，讨论一下。你认为哪种声音是噪声？

火警警报器

犬吠声

笛子演奏的音乐

声音的**响度**有大有小。响度指的是声音的**大**或**小**。有些声音可以被人们轻易地听见。这样的声音是响度较大的声音。另外一些声音则不容易被人们听见，因为它们的响度较小。

吉他弹奏的音乐

2a 什么时候你会发出响度较大的声音？
2b 什么时候你会发出响度较小的声音？

哨子声

A hand bell

A bird singing

Someone singing

We can sort sounds in different ways.

> 3a In a group, describe the different sounds made by the things in the diagrams on these pages.
>
> 3b Name one of the things that makes a high sound.
>
> 3c Name one of the things that makes a low sound.
>
> 4 Find two different ways of sorting the sounds made by these eight sound sources.

我们可以用不同的方式对声音进行分类。

3a 和你的小伙伴一起，描述一下上述图片中不同声源发出的不同声音。

3b 说出一种声音音调较**高**的声源。

3c 说出一种声音音调较**低**的声源。

4 找到两种分类方法，对这一小节中出现的八种声源进行分类。

Making Sounds

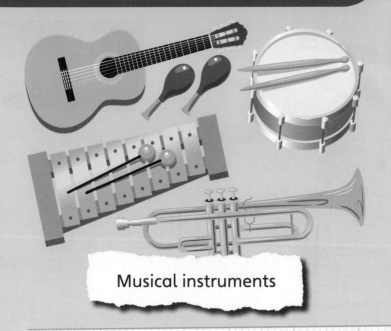

Musical instruments

Word Box
vibrating

Making sounds using instruments

You will need some instruments like the ones in the diagram.

1 Work in a group and use the instruments to make sounds.
2 Discuss how the sounds were made.

Sounds are made because something is **vibrating**. Vibrating means moving backwards and forwards very quickly. Sometimes these vibrations are easy to see, like the strings on a guitar. Sometimes these vibrations are not as easy to see, like when a whistle is blown.

📖 What vibrates to make sound on each type of musical instrument you used?

发出声音

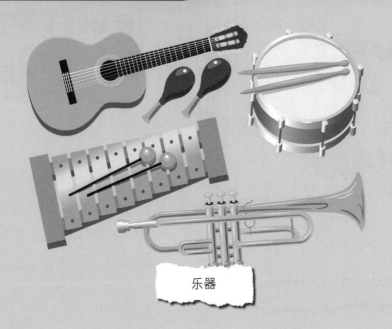

乐器

单词表
vibrating 振动

用乐器发出声音

你将需要如上图所示的一些乐器。
1 和你的小伙伴一起，演奏乐器。
2 讨论一下这些声音是如何被发出的。

声音之所以能被发出，是因为物体产生了**振动**。振动是指快速地往复运动。有时，我们很容易就能看到振动的物体，例如，吉他上的弦。但有些时候，却不太容易看到振动的物体，例如，吹哨子。

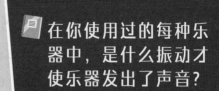
在你使用过的每种乐器中，是什么振动才使乐器发出了声音？

Vibrating Air

We know when bees are flying close to us because of the buzzing sound they make. Bees flap their wings very quickly when they move. This makes the air around them vibrate. The vibrations of the wings and the air make the buzzing sound.

1 Find out how we can hear hummingbirds hum.
2 List three other animals that make sounds as they move and explain how the sounds are made.
3 Why is it important for some animals to move very quietly?

When the guitar string is plucked it vibrates. This movement makes the air next to the string vibrate too. The air passes the vibrations to our ears and then we hear the sound.

4 Explain how we hear a whistle when it is blown.

振动的空气

我们能知道蜜蜂在离我们很近的地方飞舞,是因为它们发出了嗡嗡声。当蜜蜂移动时,会快速地扇动翅膀。这让它们周围的空气发生了振动。蜜蜂扇动翅膀引起周围空气的振动,才发出了嗡嗡声。

1 请指出我们是如何听见蜂鸟的嗡鸣声的?
2 列出其他三种在移动中发出声音的动物,并解释这些声音是如何发出的。
3 为什么对于某些动物来说,在移动时发出很小的声音至关重要?

当我们拨动吉他弦时,弦会产生振动,从而使弦附近的空气也产生振动。空气的振动传到我们的耳朵里,我们就能听见声音了。

4 请解释一下我们是如何听见哨子发出的声音的。

Explaining How We Hear

To hear sounds from the sound source we need vibrations to pass from the air to our ears. First they reach the **outer ear** and then go inside the ear until they reach the **auditory nerve**, where the vibrations are turned into signals that are sent to the brain.

Word Box
auditory nerve
outer ear

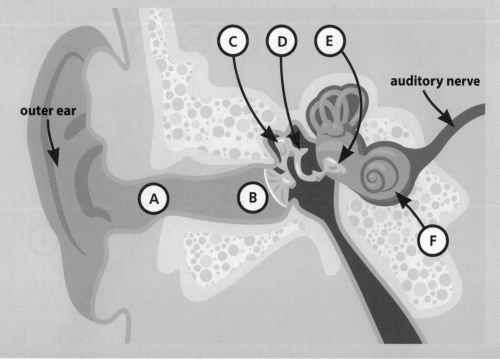

Look closely at this diagram of the parts inside our ears.
Find out what the different parts are called and how they transmit the sound.
1 Label the parts of the ear.
2 Research and explain how sound is transmitted from the outer ear to the brain.

解释我们是如何听见声音的

为了听见声源发出的声音，振动必须经由空气传递到我们的耳朵里。首先，振动会到达**外耳**，然后传入耳朵内部，到达**听觉神经**。听觉神经会将振动转化为信号，并将信号发送给我们的大脑。

单词表
auditory nerve 听觉神经
outer ear 外耳

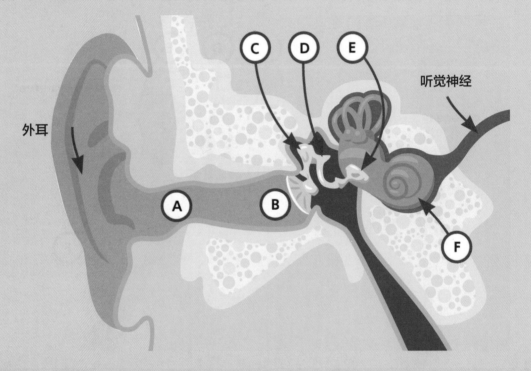

仔细观察我们耳朵内部结构的图片。
确定耳朵不同部位的名称，以及它们是如何传递声音的。
1 标注耳朵的不同部位。
2 探究并解释，声音是如何从外耳传递到我们的大脑中的。

Modelling How Sound Travels

Scientists use models to help them explain how different phenomena occur. A **phenomenon** is something we can observe as it happens but we cannot see how it is happening.

Word Box
compression wave
phenomenon

Sound is a phenomenon as we can hear the vibrations when they reach our ear, but we cannot see how they get to our ears.

Sound travels through the air as **compression waves**. We can use a 'slinky' spring as a model to help us understand these sound waves.

Slinky model

You will need: a slinky spring.

1 Work in a pair and stretch the slinky spring out like the one in the diagram.
2 Hold one end still.
3 Move the other end of the slinky quickly towards and away from the other end of the slinky by about 10 cm several times.
4 Observe what happens along the length of the slinky. This is a model for a sound wave.

模拟声音传播的过程

科学家们利用模型来帮助他们解释不同的**现象**是如何发生的。我们能观察到现象,但却观察不到它是如何发生的。

当振动传递到耳朵时,我们听到的声音就是一种现象,但是我们却不清楚振动是如何传递到我们的耳朵的。

声音是以**纵波**的形式在空气中传播的。我们可以用魔术弹簧作为模型,来帮助我们理解声波。

单词表

compression wave 纵波
phenomenon 现象

魔术弹簧模型

你将需要:一个魔术弹簧。

1. 和你的小伙伴一起,将魔术弹簧以图中所示的方式拉开。
2. 将一端固定。
3. 将弹簧另一端挤压到离固定一端大约10厘米的位置,再迅速移开,重复数次。
4. 观察弹簧的长度变化。这就是声波的模型。

Science Skills

Predict it!

Sound travels to reach our ears. If you close the door and window in your bedroom you can usually still hear some sounds from outside the room.

1a Which sounds can you sometimes hear from outside when your classroom door and windows are closed?

1b How do you hear sounds from sound sources that are outside the room?

Can sound travel through solids?

Work in a small group. You will need samples of the materials in the diagram that are larger than your hand.

1 Do you think sound will travel through each of these solids?

2 Place the sample on your table and put your ear on it. Tap the sample with your finger.

3 Does the sound travel through the material?

科学技能

预测一下吧！

声音传播到我们的耳朵。如果关上卧室的门和窗，你通常还是能听到室外的一些声音。

1a 当你关上教室的门、窗时，你偶尔能听到教室外的哪种声音？

1b 你是如何听到室外的声源发出的声音的？

声音能通过固体传播吗？

和你的小伙伴一起行动。你将用到上图中的材料样本，这些材料样本的大小要比你的手掌更大。

1 你认为声音会通过这些固体传播吗？

2 将样本放在你的桌子上，把耳朵贴到桌子上。再用你的手指轻敲样本。

3 声音能通过这种材料传播吗？

Sound Travels through Liquids

When we put our ears under water we can still hear sounds.

📕 **1a** Have you ever swum under water? What could you hear?

📕 **1b** Did the sounds sound the same as when you were out of the water?

Comparing solid, liquid and gas

Work in a small group. You will need: 3 identical plastic boxes with lids, water, sand...

1 Fill one box with water, another with sand and leave air in the third. Put the lids on tightly.

2 Place them on the table. Put your ear next to each box in turn and tap it with your finger.

3 What happens to the sound of the tapping each time?

4 Does sound travel through the solid, liquid and gas? How do you know?

声音通过液体传播

当我们在水下时,我们也能听到声音。

📄 **1a** 你曾经在水下游过泳吗?你能听到什么?

📄 **1b** 在水面下听见的声音和在水面上听见的声音是一样的吗?

比较固体、液体和气体

和你的小伙伴一起行动。你将需要:三个一模一样的带盖子的塑料盒、水、沙子……

1 分别用水和沙子装满两个塑料盒,另外一个什么也不装。把三个塑料盒的盖子盖紧。

2 把塑料盒放到桌子上。再把耳朵依次贴近每一个塑料盒,用手轻敲塑料盒。

3 每次轻敲发出的声音听起来是怎样的?

4 声音能通过固体、液体、气体传播吗?你是如何知道的?

Changing Pitch

Every sound has a **frequency**. The frequency is the number of vibrations in one second. This determines how high or low a sound is. We call this the **pitch** of the sound. A high sound is squeaky like a mouse. A low sound is deep like the bark of a big dog.

> **Word Box**
> frequency
> pitch

A referee's whistle makes very fast vibrations so it produces many vibrations in a second and a very high-pitched sound.

A double bass makes lower-pitched sounds than a whistle. The vibrations it makes are slower so it produces fewer vibrations in a second and a low sound.

> With a partner, discuss sound sources that make high and low sounds.
> 1a List three sound sources that produce high sounds.
> 1b List three sound sources that produce low sounds.

变化的音调

每种声音都有**频率**。频率是指每秒振动的次数。它决定了声音的高低。我们把声音的高低叫作**音调**。频率较高的声音是较为尖厉的声音,例如,老鼠发出的吱吱声。频率较低的声音是较为低沉的声音,例如,犬吠声。

单词表
frequency 频率
pitch 音调

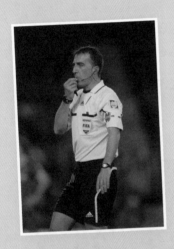

裁判的哨子能产生很快的振动,因此它每秒振动次数很多,音调就会非常高。

低音提琴的音调比哨子的音调低。它的振动频率更慢,因此它在单位时间里的振动次数较少,音调较低。

和你的小伙伴一起,讨论一下,能发出音调高的和音调低的声音的声源。

 1a 列出三个能发出音调较高的声音的声源。

 1b 列出三个能发出音调较低的声音的声源。

Changing sounds

Look at this picture of instruments in an orchestra. They all sound different. How are the sounds made? How can the musicians change the sounds of their instruments? What can they do to alter the volume and pitch of a note?

Sensing sounds

All mammals, like us, have ears. We use our ears to detect sounds. Some animals have huge ears which are very sensitive and collect a lot of sounds.

Dolphins use sound echoes to find their way around.

The fennec fox can hear sounds which are too quiet for humans to hear.

改变声音

观察这张照片中管弦乐团的乐器。每种乐器的声音都不一样。这些声音都是如何被发出来的？这些音乐家都是如何改变乐器的声音的？他们要怎么做才能改变音符的响度和音调？

感受声音

所有的哺乳动物像人类一样都有耳朵。我们用耳朵来探测声音。一些动物的耳朵很大，十分敏感，能听到很多声音。

海豚可以通过回声找到方向。

耳廓狐能听见人类听不见的声音。

Did you know?

An echo happens when a sound wave bounces off objects and returns to our ears a moment after the original sound.

Find out

find out how bats use sound to find their way around.

Secret sounds

Some sound is too high-pitched for humans to hear. We call this ultrasound. Doctors send waves of ultrasound into a body. The sound bounces off our organs and can create a picture of what is inside us.

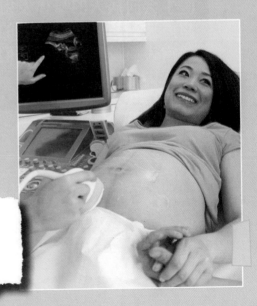

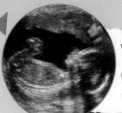

We can see a baby growing before it is born using ultrasound.

你知道吗？

声波在传出去以后，在物体表面反弹，经过一段时间后，我们听到的不同于原始声音的声波就是回声。

发现一下

去探索蝙蝠是如何利用声波确定方向的。

神秘的声音

一些声音的频率太高以至于人们听不见。我们把这种声音叫作超声波。医生会向病人的身体发射超声波。超声波在我们的器官表面反射，生成我们身体内部的图像。

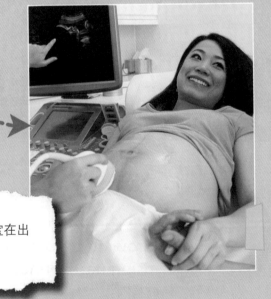

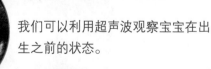

我们可以利用超声波观察宝宝在出生之前的状态。

Exploring Changing Pitch

Different musical instruments have different frequencies of vibrations, so make sounds of different pitches. We bang, shake, pluck or blow instruments to create vibrations and make sounds.

High and low sounds

You will need: instruments similar to those in the picture.

1. Work in a small group. Play the instruments and discuss the pitches of the sounds they produce.
2a. Classify the instruments into those that make a high sound, those that make a low sound and those that can make sounds at various pitches.
2b. Record your findings on a Venn diagram.
3. Name another instrument that could be added to each group.

探究变化的音调

不同的乐器有不同的振动频率，因此会发出不同音调的声音。我们重击、抖动、弹拨或吹奏各种乐器来产生振动，发出声音。

高音和低音

你将需要：与上图类似的各种乐器。

1. 和你的小伙伴一起，分别演奏各种乐器并讨论乐器发出的声音的音调高低。
2. a 将乐器分为产生音调高的乐器、产生音调低的乐器和产生不同音调的乐器。
2. b 用文氏图记录你的发现。
3. 想出另外三种可以分别归为上述三类的乐器。

Science Skills

Investigate it!

We can make our own instruments that can make sounds at different pitches using everyday objects. High-pitched sounds are short, fast (high-frequency) vibrations. Low-pitched sounds are longer, slower vibrations (low-frequency).

Making instruments

You will need: a set of baby stacking cups (or plastic food boxes of different sizes without lids), rubber bands of different lengths and thicknesses... Work in a small group.

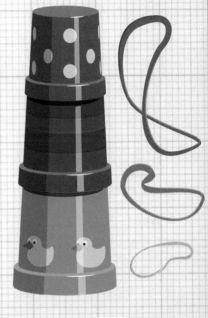

1a Explore making sounds by stretching the same size of elastic band over the different-sized stacking cups and plucking the band.

1b How does the pitch differ over the different-sized cups? Why?

2 Predict how changing the thickness of the band changes the pitch of the sound.

科 学 技 能

探究一下吧！

我们可以用日常物品来自己制作音调高低不同的乐器。音调较高的声音由较短、较快（高频率）的振动产生。音调较低的声音由较长、较慢（低频率）的振动产生。

制作乐器

你将需要：一组儿童叠杯（或不同大小的无盖塑料食物盒）、不同长度和不同厚度的橡皮筋……

和你的小伙伴一起。

1a 将相同的橡皮筋放在不同大小的叠杯上方并拉直，拨动橡皮筋，听发出的声音。

1b 不同大小的杯子的音调有什么不同？为什么？

2 请预测改变橡皮筋的厚度会如何影响音调。

3 How could you carry out an investigation to find out whether changing the thickness of the band changes the pitch of the sound?

4 Investigate to test your predictions. Discuss your results.

Oboe straws

You will need: plastic drinking straws, a pair of scissors...

1. Squash and cut the end of the straw like the one in the diagram.
2. Place the straw about 2 cm into your mouth and close your lips around the straw.
3. Blow until you make a sound.
4. How do you think you can make oboe straws that make sounds at different pitches? Test to find out.

5a How did the pitch change with length of the oboe straw?

5b Why did it change in this way?

3 你能设计实验来探究橡皮筋厚度是如何影响声音的音调的吗？

4 通过实验来检验你的预测是否正确。讨论一下你的结果。

吸管双簧管

你将需要：塑料吸管、一把剪刀……

1 以如图所示的方式压扁并剪开吸管的一端。

2 将吸管放进嘴里约2厘米，上下嘴唇闭合。

3 试着吹出声音。

4 你认为怎样做才能制作出发出不同音调的声音的吸管双簧管？检验一下。

5a 音调会如何根据吸管双簧管的长度而发生变化？

5b 为什么会发生这种变化？

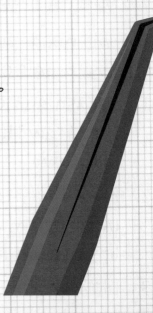

Measuring the Volume of Sounds

Sounds can be loud or quiet. The loudness of a sound is measured in **decibels** (dB). We call how loud a sound is its volume.

Word Box
decibel

Large vibrations make loud sounds. Small vibrations make quieter sounds. The vibrations of this pneumatic drill produce sounds that are loud enough to damage human hearing. Workers in very loud environments need sound-absorbing ear defenders to protect their hearing.

Changing the volume of sounds

You will need: a variety of different types of instruments, like the ones on page 25.

1. Play the instruments. Explore how to make loud and quiet sounds with the instruments.
2. How did you change the volume of the sound?

How is the sound volume different around your school? Use a data logger to measure volume in different places and compare the volumes.

测量声音的响度

声音可以洪亮也可以微弱。声音的响度可以用**分贝（dB）**来衡量。我们把声音的响度叫作音量。

单词表
decibel 分贝

幅度大的振动能产生响度较大的声音。幅度小的振动能产生响度较小的声音。风钻振动发出的声音的响度足够大，导致声音能损害人类的听力。在喧闹的环境工作的人们需要佩戴能吸收声音的防护工具来保护听力。

改变音量的大小

你将需要：如25页图所示的各种各样的乐器。

1. 演奏乐器。探究一下如何能让乐器分别发出响度较大的和响度较小的声音。
2. 你是如何改变音量大小的？

你所在的学校周围各种声音的响度有什么不同？用一个数据记录器来测量不同地方的声音的音量，并进行比较。

Science Skills

Far Away Sounds – Plan it!

Have you ever waited for a train? Sometimes you can hear a train approaching a station before you see it.

1 What happens to the sound of a train as it gets closer?

2 Do you think this fennec fox can hear sounds that are too far away for us to hear? Why?

Sometimes we are too far away to hear sounds. Some children say this is why they cannot hear the bell being rung at the end of play time.

3 In your group discuss how you could do an investigation to find out how far away you would need to be before you could not hear a small bell ringing.
4 Record how you would find out.
5 Carry out your investigation and record your results.

科学技能

远处的声音——做个计划吧！

你等过火车吗？有时你还没看到火车就能听到它靠近站台了。

1 火车在靠近时，它发出的声音是怎样的？

2 你认为耳廓狐能听见那些距离人们太远而人们无法听到的声音吗？为什么？

有时我们距离声源太远而无法听到声音。有些小朋友说这就是他们在休息时间结束的时候听不到铃声的原因。

3 和你的小伙伴一起，讨论一下，你们要如何探究你能听到铃声的最远距离。

4 记录你们的探究方式。

5 进行实验并记录你们的结果。

What Do We Know About Sound?

1 List three sounds you can hear and identify the sound source.

1 _____ made by _____

2 _____ made by _____

3 _____ made by _____

2 Describe the sounds that each of the following sound sources make.

A baby _____

A car _____

A keyboard _____

3 How do you think you can hear sounds from a sound source that is not next to you?

_____
_____
_____

Classifying Sounds

Find two different ways of sorting (classifying) the sounds made by these sound sources. Say how you have classified them. Think about volume and pitch. You can use these words to help you:

loud, quiet, low, high, noise, variety, blow, bang, pluck, natural

Fire alarm Guitar Whistle Dog

Flute Bird Hand bell Singer

One way of classifying these sounds is:

Another way of classifying these sounds is:

Making Sounds

1. What is vibrating with these instruments?

Instrument	Part of instrument vibrating
Guitar	
Drum	
Rainmaker	
Violin	
Xylophone	
Cymbal	
Ukulele	
Maracas	
Castanets	

2. We can make a musical instrument out of recycled materials. How would you make a musical instrument that vibrates to make sound? Draw and label your idea.

Vibrating Air

1. Find out how we can hear hummingbirds hum.

2. List three other animals that make sounds as they move and explain what vibrates to make the sound.

Animal	What vibrates to make the sound

3. Why is it important for some animals to move quietly?

4. What happens when this whistle is blown that means we can hear it?

Science Skills

Investigate it!

1 Predict whether sound will travel through these solid materials. Record your predictions in this table.

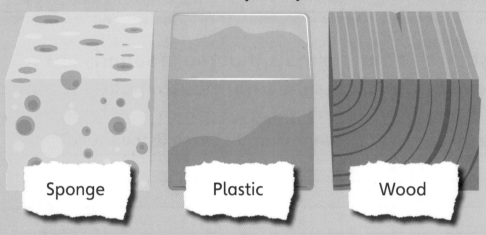

Material	Will sound travel through it? Yes/No
Wood	
Metal	
Plastic	
Cardboard	
Rubber	
Dough	
Sponge	
Cotton wool	

We can test to find out whether sound travels through solids by placing our ear on them and tapping them with our finger.

2a What will you hear if the solid material is letting sound travel through it?

_____

2b Record your results in this table.

Material	Did sound travel through it? Yes/No
Wood	
Metal	
Plastic	
Cardboard	
Rubber	
Dough	
Sponge	
Cotton wool	

3 Did your results match your predictions? Why?

_____

_____

Explaining How We Hear

1. Label the parts of the ear.

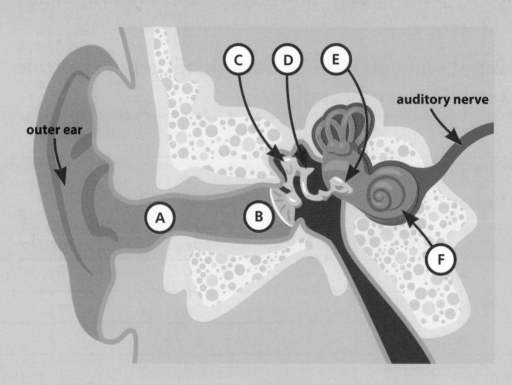

A		D	
B		E	
C		F	

2. Explain how sound is transferred from the outer ear to the brain.

_____
_____
_____
_____

Sound Travels through Liquids

1. Have you ever swum under water? What could you hear?

2. How were the sounds different from when you were out of the water?

3. Find out how dolphins communicate with each other. Record your findings.

Changing Pitch

1. List three sound sources that produce high sounds.

   1. _____
   2. _____
   3. _____

2. List three sound sources that produce low sounds.

   1. _____
   2. _____
   3. _____

3. Why is it nice to be able to press keys that each play a different pitch on a piano?

   _____
   _____

Exploring Changing Pitch

1. Classify these instruments on the Venn diagram into those that produce high-pitched sounds, those that produce low-pitched sounds and those that can produce sounds at various pitches.

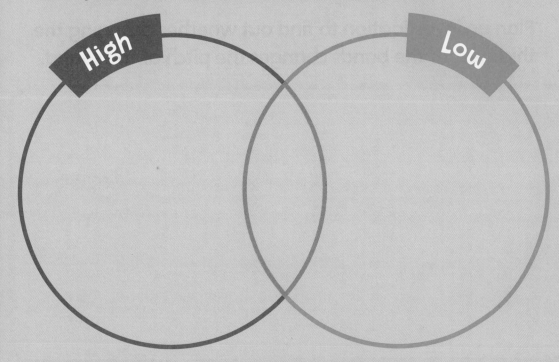

2. Name one other instrument that can be added to each group and add them to the diagram.

Science Skills

Investigate it!

This diagram shows how we can make a string instrument using a box and rubber bands.

1   Plan an investigation to find out whether changing the thickness of the bands changes the pitch of the sound.

2   What do you predict?

You will need three hexagonal nuts and three round balloons. Put each hexagonal nut into a balloon and blow up the balloon. Tie it. Hold the knot and rotate the balloon. Listen carefully to the sound it makes.

3   Predict whether the pitch will be different if you rotate each balloon at the same speed? Why? Record your prediction.

I predict _____
_____
_____

4   Now rotate each balloon. How did the sound compare to your prediction? Record your results.

I found out that _____
_____
_____

Measuring the Volume of Sounds

1 Describe how you can make loud and quiet sounds with these musical instruments.

Instrument	Loud	Quiet
Drum		
Whistle		
Rattle		
Harp		

2 Draw a map of part of your school. Label places that are usually quiet and places that are sometimes loud.

Science Skills

Plan it!

1. How would you carry out an investigation to find out how far away you would need to be before you could no longer hear a small bell ring?

   _____
   _____
   _____
   _____

2. This graph shows how far away four different children were before they could not hear the bell. List three things the graph shows.

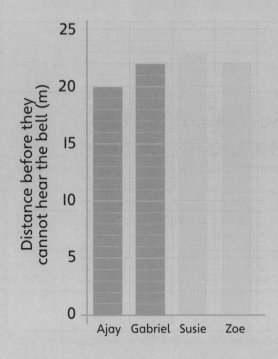

1 _____

2 _____

3 _____

What Do We Know?

1. Research 'Foley artists' (people who make the sound effects for a film, television or radio) then choose three different instruments that make sounds in different ways. Describe two ways each one could make different sounds.

Instrument	Ways to change the sound it makes

2. Record everything you know about the link between vibrations and the sounds we hear.

_____
_____
_____
_____

Glossary

auditory nerve: the nerve at the back of the ear that tells the brain what has been heard

compression wave: the type of wave that sound is

decibel: unit we use to measure how loud sounds are

frequency: number of sound vibrations in a second

high (sound): squeaky sound (like a mouse makes); high-pitched

loud: sound that is very easy to hear; if a sound is very loud it can hurt our ears

low (sound): gruff sound (like a bear makes); low-pitched

noise: sound that can be unpleasant

outer ear: visible part of the ear on the side of our head

phenomenon: something we can observe happening but not observe *how* it is happening

pitch: how high or low a sound is

quiet: sound that is not easy to hear

source (sound): the thing that makes the sound

vibrating: moving backwards and forwards very quickly

volume: how quiet or loud a sound is

词汇表

听觉神经：耳朵内部向大脑传递信息的神经

纵波：和声波相似的波

分贝：用来衡量声音大小的单位

频率：声音每秒振动的次数

高的（与声音有关的）：尖锐的声音，音调高（例如，老鼠发出的叫声）

大声的：容易听到的声音，如果声音的音量过大，可能会损害我们的耳朵

低的（与声音有关的）：低沉的声音，音调低（例如，熊发出的声音）

噪声：令人难受的声音

外耳：头两侧露在外面的耳朵部分

现象：我们能观察到的事物，但无法观察到它是如何发生的

音调：形容声音的高低

小声的：不容易被听到的声音

来源（与声音有关的）：发出声音的物体

振动：快速地前后移动

响度：形容声音的大小

培生科学虫双语百科
奇妙物理

Forces

力

英国培生教育出版集团 著·绘
徐昂 译

电子工业出版社
Publishing House of Electronics Industry
北京·BEIJING

Original edition, entitled SCIENCE BUG and the title Forces Topic Book, by Tanya Shields published by Pearson Education Limited © Pearson Education Limited 2018

ISBN: 9780435195847

All rights reserved. No part of this book may be reproduced or transmitted in any form or by any means, electronic or mechanical, including photocopying, recording or by any information storage retrieval system, without permission from Pearson Education Limited.

This adaptation of SCIENCE BUG is published by arrangement with Pearson Education Limited.

Chinese Simplified Characters and English language (Bi-lingual form) edition published by PUBLISHING HOUSE OF ELECTRONICS INDUSTRY, Copyright © 2023.

For sale and distribution in the mainland of China exclusively (except Hong Kong SAR, Macau SAR and Taiwan).

本书中英双语版由Pearson Education（培生教育出版集团）授权电子工业出版社在中华人民共和国境内（不包括香港、澳门特别行政区及台湾地区）独家出版发行。未经出版者书面许可，不得以任何方式抄袭、复制或节录本书中的任何部分。

本套书封底贴有Pearson Education（培生教育出版集团）激光防伪标签，无标签者不得销售。

版权贸易合同登记号　图字：01-2022-2381

图书在版编目（CIP）数据

培生科学虫双语百科. 奇妙物理. 力：英汉对照 / 英国培生教育出版集团著、绘；徐昂译. --北京：电子工业出版社，2024.1

ISBN 978-7-121-45132-4

Ⅰ.①培… Ⅱ.①英… ②徐… Ⅲ.①科学知识－少儿读物－英、汉 ②力学－少儿读物－英、汉 Ⅳ.①Z228.1 ②O3-49

中国国家版本馆CIP数据核字（2023）第035079号

责任编辑：李黎明　文字编辑：王佳宇
印　　刷：河北迅捷佳彩印刷有限公司
装　　订：河北迅捷佳彩印刷有限公司
出版发行：电子工业出版社
　　　　　北京市海淀区万寿路173信箱　邮编：100036
开　　本：787×1092　1/16　印张：35　字数：840千字
版　　次：2024年1月第1版
印　　次：2024年2月第2次印刷
定　　价：199.00元（全9册）

凡所购买电子工业出版社图书有缺损问题，请向购买书店调换。若书店售缺，请与本社发行部联系，联系及邮购电话：（010）88254888，88258888。

质量投诉请发邮件至zlts@phei.com.cn，盗版侵权举报请发邮件至dbqq@phei.com.cn。
本书咨询联系方式：010-88254417，lilm@phei.com.cn。

使用说明

　　欢迎来到少年智双语馆！《培生科学虫双语百科》是一套知识全面、妙趣横生的儿童科普丛书，由英国培生教育出版集团组织英国中小学科学教师和教研专家团队编写，根据英国国家课程标准精心设计，可准确对标国内义务教育科学课程标准（2022年版）。丛书涉及物理、化学、生物、地理等学科，主要面向小学1~6年级，能够点燃孩子对科学知识和大千世界的好奇心，激发孩子丰富的想象力。

　　本书主要内容是小学阶段孩子需要掌握的物理知识，含9个分册，每个分册围绕一个主题进行讲解和练习。每个分册分为三章。第一章是"科学虫趣味课堂"，这一章将为孩子介绍科学知识，培养科学技能，不仅包含单词表、问题和反思模块，还收录了多种有趣、易操作的科学实验和动手活动，有利于培养孩子的科学思维。第二章是"科学虫大闯关"，这一章是根据第一章的知识点设置的学习任务和拓展练习，能够帮助孩子及时巩固知识点，准确评估自己对知识的掌握程度。第三章是"科学词汇加油站"，这一章将全书涉及的重点科学词汇进行了梳理和总结，方便孩子理解和记忆科学词汇。

　　2024年，《培生科学虫双语百科》系列双语版由我社首次引进出版。为了帮助青少年读者进行高效的独立阅读，并方便家长进行阅读指导或亲子共读，我们为本书设置了以下内容。

　　（1）每个分册第一部分的英语原文（奇数页）后均配有对应的译文（偶数页），跨页部分除外。读者既可以进行汉英对照阅读，也可以进行单语种独立阅读。问题前面的 📖 符号表示该问题可在第二部分预留的位置作答。

　　（2）每个分册第二部分的电子版译文可在目录页扫码获取。

　　（3）本书还配有英音朗读音频和科学活动双语视频，也可在目录页扫码获取。

　　最后，祝愿每位读者都能够享受双语阅读，在汲取科学知识的同时，看见更大的世界，成为更好的自己！

电子工业出版社青少年教育分社
2024年1月

Contents 目录

Part 1　科学虫趣味课堂　　　　　　　　　　　　　　　/ 1

Part 2　科学虫大闯关　　　　　　　　　　　　　　　　/ 47

Part 3　科学词汇加油站　　　　　　　　　　　　　　　/ 71

Part2译文

配套音视频

Forces

Forces make things **move**. Forces can be pushes or pulls, they can make things move or make them stop. They can slow things down, speed them up and make them change direction too.

Word Box
force
move

Look at the diagram.
1 How many moving things can you name?
2 How do you make these things move?
3 How do you make these things slow down?

力

力能让物体**移动**。力可以是推力或拉力，这些力能让物体移动或停止。力可以减慢或加快物体运动的速度，也能改变物体运动的方向。

单词表
force 力
move 移动

观察图片
1 你能说出多少个正在移动的物体？
2 如何让这些物体移动？
3 如何让这些物体慢下来？

What Forces Do

To make something move a force is needed. There are two main types of force: pushes and pulls. When you open a drawer you **pull** the handle and the drawer moves towards you. To close the drawer you **push** the handle and the drawer moves away from you.

Word Box
pull
push

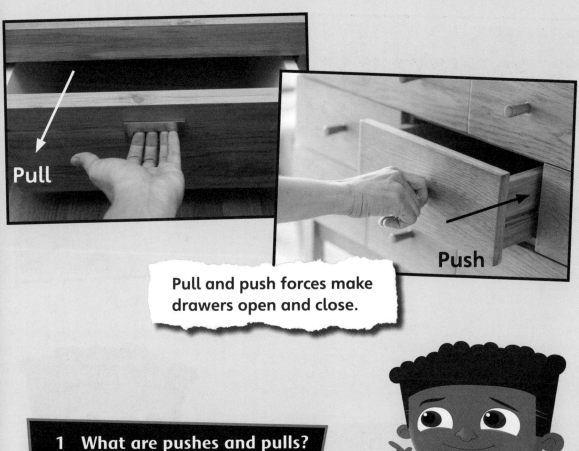

Pull

Push

Pull and push forces make drawers open and close.

1. What are pushes and pulls?
2a List three things you have moved today.
2b How did you make them move?

力的作用

为了让物体移动,人们需要使用力。力主要分为两类:推力和拉力。当你打开抽屉时,**拉**动把手,可以将抽屉拉向自己。为了关上抽屉,**推**动把手,抽屉就远离自己。

> **单词表**
> pull 拉;拉力
> push 推;推力

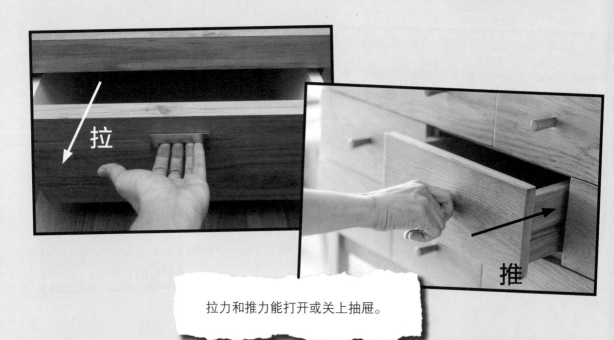

拉力和推力能打开或关上抽屉。

1 什么是拉力和推力?
2a 请列出你今天移动过的三件物品。
2b 你是如何移动它们的?

Forces and sports

Athletes use forces to move objects or move themselves. A sprinter uses both legs to push against the starting blocks at the start of a race. An archer pulls the string on their bow before firing arrows at a target.

A rider pulls on the reins to control the horse.

A footballer kicks the ball and pushes it forward.

- 3a Make a list of sports that could be in a sports competition.
- 3b Write about how each sport uses forces to make things move.

力与体育运动

运动员通过力的作用使物体移动或让自己移动。短跑运动员会在起跑时用两条腿蹬助跑器，对自己施加推力。射箭运动员会在射击前拉满弓弦，施加拉力。

马术运动员拉缰绳，产生拉力来控制马。

足球运动员踢球，产生推力让足球向前运动。

- 3a 请列出运动会上可能出现的各种体育运动。
- 3b 请写出每种运动是如何通过力的作用让物体移动的。

Sir Isaac Newton

Sir Isaac Newton was one of the most important scientists in history. He was born in England on the 25th December 1642 and lived until he was 85.

Word Box
gravity

When he was 19, Newton went to study mathematics at the University of Cambridge. This is where he spent most of his life. He liked to work alone, never married and did not have many friends. He was made a knight by Queen Anne in 1705, the first scientist to receive this honour. This is why we call him Sir Isaac Newton.

Use your research skills to discover more about Sir Isaac Newton.

1. What were his main discoveries?
2. Make a list of the facts you find interesting.

艾萨克·牛顿爵士

艾萨克·牛顿爵士是历史上最具影响力的科学家之一。1642年12月25日他出生于英国，享年85岁。

单词表
gravity 引力；地心引力

在19岁时，牛顿赴剑桥大学学习数学。他在剑桥大学度过了一生中大部分的时光。他喜欢独自工作，终身未婚，也没有太多朋友。1705年，英国女王安妮授予牛顿"爵士"的称号，他是第一个获此殊荣的科学家。这就是我们叫他艾萨克·牛顿爵士的原因。

运用你的调查技能，找出更多关于艾萨克·牛顿爵士的信息。

1. 他的主要研究成果有哪些？
2. 请列出与他相关的你认为的有意思的事。

Discovering gravity

Isaac Newton is best known for discovering **gravity**. The story is that one day while sitting near an apple tree he watched an apple fall from a branch. He realised that some force must be making the apple fall towards the ground or it would have stayed where it was on the tree.

He named this force gravity.

In 1687 Newton wrote down his scientific ideas and published a book called *Principia*. This book is one of the most important scientific books ever published. It not only explained the idea of gravity, but also described how all things move.

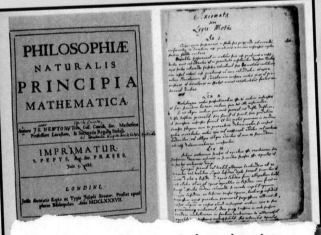

You can see Newton's hand-written corrections for the second edition of his book.

3 Use your research on Sir Isaac Newton to create a mini-book about him. Don't forget to include interesting facts about his work.

发现地心引力

艾萨克·牛顿是以发现**地心引力**而闻名的。相传,有一天,他坐在苹果树旁,看到树枝上的一颗苹果掉下来。于是他意识到,一定是有某种力让苹果掉下来,否则苹果就会一直留在树上。

他将这种力命名为地心引力。

1687年,牛顿记录了自己的科学思想,并出版了书籍《自然哲学与数学原理》。这本书成了科学史上最重要的书籍之一。这本书不仅解释了重力的概念,还描述了物体的运动规律。

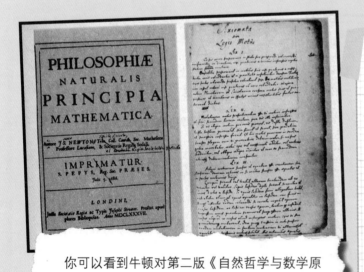

你可以看到牛顿对第二版《自然哲学与数学原理》进行的手写修改。

3 请你将关于艾萨克·牛顿爵士的调研结果制作成一本迷你书。别忘了把关于他研究的有意思的事情也写进去。

Newton Meters

A force meter, also called a **newton meter,** is used to measure forces. The **newton** or N for short is the unit of force and is named after Sir Isaac Newton.

A newton meter has a spring with a hook attached to it. The spring stretches when a force is applied to the hook. As the spring stretches, the force is measured on the scale.

Word Box
newton (N)
newton meter

spring

scale

hook

弹簧测力计

测力计，也被称为**弹簧测力计**，可以用来测量力的大小。**牛顿**（或缩写为N）是力的单位，是以艾萨克·牛顿爵士的名字命名的。

弹簧测力计包括一根弹簧和一个附在弹簧上的钩子。当力施加在钩子上时，弹簧就会伸长。当弹簧伸长时，标尺上的刻度就会显示力的大小。

单词表

newton (N) 牛顿
newton meter 弹簧测力计

弹簧

刻度

钩子

Some newton meters are constructed so they can measure small forces and some are made to measure larger forces. The stronger the spring inside the meter the larger the force it can measure.

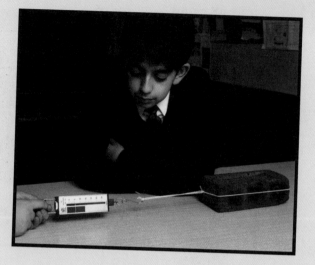

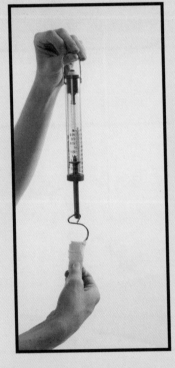

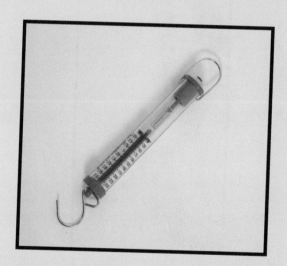

On Earth a force of 10 N is needed to lift a 1 kg mass.

> Use a newton meter to find what force is needed to lift some everyday objects. Try a chair or a shoe.

一些弹簧测力计能用来测量较小的力，一些则可以测量较大的力。弹簧测力计中弹簧的弹性越强，能测量的力就越大。

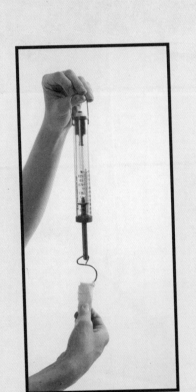

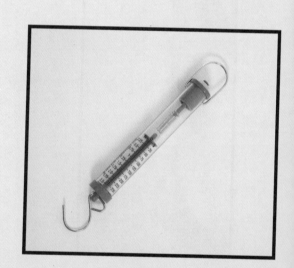

在地球上，10牛的力可以提起质量为1千克的物体。

使用弹簧测力计来测量需要多大的力来提起日常物品。拿一把椅子或一只鞋尝试一下。

Gravity

Gravity is a force that cannot be seen, but its effects can be felt in everything we do. Gravity is a pull force that makes things fall to the ground and stay there. Gravity is the force that stops everything floating off into **space**.

Word Box
Earth
space

Astronauts float in space as the pull of gravity is not as strong away from Earth.

When you jump up your legs push you off the ground. The force of gravity pulls you back down.
If gravity did not exist you could jump off the **Earth** and never fall back down.
Gravity is a pull force.

Gravity

1 List three things that would be easier if there were less gravity on Earth.

2 List three things that would be more difficult if there were less gravity on Earth.

地心引力

我们观察不到地心引力,但是却能随处感受到地心引力的作用。地心引力是一种拉力,能让物体落向地球表面并停留在地球表面。地心引力是阻止地球表面上的物体飘向**太空**的一种力。

单词表
Earth 地球
space 太空

宇航员能漂浮在太空中,这是因为太空中地心引力的作用没有地球上的大。

当你跳起来时,你的双腿将你推离地面。而地心引力则施加拉力将你拉回地面。

如果没有地心引力,你跳起来就会离开**地球**,而且不会再落回**地球**。

地心引力是一种拉力。

1 如果地球上的地心引力变小,请列出三件人们更容易做到的事。

2 如果地球上的地心引力变小,请列出三件人们更难做到的事。

地心引力

Gravity on Other Planets

The force of gravity also keeps the planets in **orbit** around the Sun. All massive objects such as the Earth, Sun and Moon exert a force that pulls things towards them, called a **gravitational pull**.
Gravity keeps the Earth at a fixed distance from the Sun (the distance varies by only 4% throughout the year).
The gravitational pull between the Earth, Sun and Moon keeps the Earth in a constant orbit around the Sun, and the Moon in constant orbit around the Earth.

Word Box
gravitational pull
orbit
tide

Gravity on the Moon

The Moon's gravitational pull is weaker than the Earth's as it has less mass, but we can still see its effect on our oceans.
Ocean **tides** go in and out as the Moon orbits the Earth and the Moon's gravitational pull moves water towards it.

The gravitational pull of the Sun keeps each planet in orbit around it.

In a pair discuss:
What would life be like if we did not have a Moon?

The gravitational pull of the Moon is one-sixth of the Earth's gravitational pull.

其他行星上的引力

引力也会让太阳周围的星球在固定**轨道**上运行。所有巨大的物体，例如，地球、太阳和月球都会对其他物体产生一种引力，即**万有引力**。

引力让地球与太阳保持固定距离（一年之中这个距离的变化幅度在4%左右）。

地球、太阳和月球之间的万有引力使地球围绕着太阳在固定的轨道上运行，而月球则围绕着地球在固定的轨道上运行。

单词表
gravitational pull 万有引力
orbit 轨道
tide 潮汐

太阳的万有引力作用让周围的行星在固定的轨道上运行。

月球上的引力

因为月球比地球的质量小，所以月球的万有引力作用比地球的更小，但我们仍然能观察到月球的引力对我们的海洋产生的作用效果。

随着月球绕着地球运行，月球的万有引力作用会对海水产生拉力，海洋出现**潮汐**。

和你的小伙伴一起，讨论一下：
如果没有月球，地球上的生活将会是怎样的？

月球的万有引力作用是地球的六分之一。

Science Skills

Gravity – Graph it!

Word Box
mass
weight

Mass and **weight** are easy to confuse. The mass of an object is a measure of how much matter it contains. For example, a book contains less matter than a table so it has less mass. An object's mass is the same wherever it is, even in space. Mass is measured in grams (g) or kilograms (kg).

Weight is a force caused by gravity. The weight of an object is the downward gravitational force between it and the Earth. On the surface of the Earth the more mass an object has the greater its weight will be. Weight is measured in newtons (N).

All objects have mass, and so exert the force of gravity. Even you attract other objects to you because of gravity, but you have too little mass for the force to be very strong. Gravity only becomes noticeable with a massive object such as a moon, planet or star.

My mass is 60 kg.

When I am on Earth my weight is 600 N.

Gravity on the Moon is 1/6th of the Earth's gravity. What would I weigh on the Moon?

1. Which planets have a greater mass than Earth?
2. Which planets have a smaller mass than Earth?

科学技能

引力——画个图表吧！

单词表
mass 质量；重物
weight 重力

人们很容易把**质量**和**重力**这两个概念混淆。一个物体的质量是指其所含物质的多少。例如，一本书所含的物质比一张桌子少，因此书的质量更小。无论在哪里，一个物体的质量都是不变的，包括在太空中。质量的单位是克（g）或千克（kg）。

重力是由于物体受到引力而产生的一种力。一个物体的重力是地球和物体之间产生的方向向下的引力。在地球表面，一个物体的质量越大，其重力就越大。重力的单位是牛顿（N）。

所有物体都有质量，因此都会产生引力。由于引力的作用，你自己甚至也会吸引其他物体，但是由于你的质量太小，这种引力的大小也就不大。引力只有在质量大的物体上才变得明显，例如，卫星、行星或恒星。

我的质量是60千克。

当我在地球上时，我的重力是600牛。

我在月球上所受的引力是在地球上的六分之一。我在月球上的重力应该是多少？

地球

1 哪些行星的质量比地球大？
2 哪些行星的质量比地球小？

How heavy would a bag of sugar feel?

A standard bag of sugar with a mass of 1 kg weighs 10 N on Earth. If we took it to a planet with less gravity it would feel lighter. If we took it to a planet with more gravity it would feel heavier.

The data in the table shows how heavy the bag of sugar would feel on each planet.

> 3 Use the data in the table below to draw a graph to show how much a bag of sugar would weigh on different planets. Remember, the mass of something is a measure of how much matter it contains. This does not change. The weight of something is a measure of the force of gravity acting on the object. This can change depending on where the object is.

Planet	Mass of sugar	Gravitational pull compared to Earth's (approx.)	Weight
Mercury	1 kg	1/3rd	3.8 N
Venus	1 kg	9/10th	9.1 N
Earth	1 kg		10 N
Mars	1 kg	1/3rd	3.8 N
Jupiter	1 kg	2.5 times stronger	23.6 N
Saturn	1 kg	Slightly stronger	10.6 N
Uranus	1 kg	4/5th	8.9 N
Neptune	1 kg	1/10th stronger	11.3 N

> 4 Which planets have a similar gravity to Earth?

一袋糖有多重？

在地球上，一袋糖的质量是1千克，重力是10牛。如果我们把这袋糖放到引力更小的行星上，人们会感觉它变轻了。如果我们把这袋糖放到引力更大的行星上，人们会感觉它变重了。

表格中的数据说明了这袋糖在不同行星上的重力是如何变化的。

3 请利用表格中的数据绘制一张曲线图，说明一袋糖在不同行星上的重力。请牢记，一个物体的质量是指其所含物质的多少，质量是不会改变的。而物体的重力则表明了行星施加在物体上的引力的大小。物体的重力会随着物体所处地点的变化而变化。

行星	糖的质量	其他行星的万有引力作用与地心引力的比值（近似值）	重力
水星	1千克	1/3	3.8牛
金星	1千克	9/10	9.1牛
地球	1千克		10牛
火星	1千克	1/3	3.8牛
木星	1千克	大于2.5倍	23.6牛
土星	1千克	稍大一点儿	10.6牛
天王星	1千克	4/5	8.9牛
海王星	1千克	大于1/10	11.3牛

4 哪些行星的引力大小与地心引力相似？

Friction

When two surfaces are in contact with each other and one or both of them are moving, or trying to move, there will always be **friction**. Friction is a force and always slows down a moving object. Even objects moving through air experience friction.

Word Box
friction

Look at the areas highlighted in the picture.
1 What two surfaces are in contact with each other in each example?
2 Where else could there be friction?

When the tyres move on the road surface friction is generated.

Rub the palms of your hands together. The quicker you rub your hands the warmer they get. As the surface of your hands slide against each other they create friction and this produces heat.

摩擦力

当两个物体的表面相互接触并产生或将要产生相对运动时，两个物体间会产生**摩擦力**。摩擦力会降低物体的运动速度，甚至，在空气中运动的物体也会产生摩擦力。

单词表
friction 摩擦力

观察图片中圈出来的部分。
1 在图片圈出来的部分中，哪两个部分的表面相互接触？
2 还有哪里可能产生摩擦力？

当轮胎在地面上移动时，会产生摩擦力。

搓一搓你的两只手，搓得越快手就会感觉越热。这是因为你的两只手表面相互接触，产生了摩擦力，从而产生了热量。

If there is enough friction it can even stop things moving, for example, applying the brakes on a bicycle increases the friction between the brake and the wheel and causes the wheel to stop rotating. The amount of friction produced is different for different materials. Usually, smooth surfaces produce less friction and rough surfaces produce more friction.

Sometimes we want to reduce friction. For example, we use oil to reduce the friction between the moving parts inside a car engine. The oil holds the surfaces apart and flows between them.

Brakes use friction to stop the wheel from rotating.

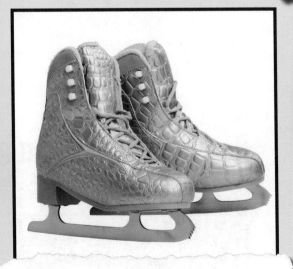

Ice skaters wear special boots that produce very little friction.

3 Make a list of the different types of shoe that use friction to stop the wearer from slipping.
4 Which sports need equipment that produces very little friction?

如果摩擦力足够大，它还能让物体停止运动，例如，自行车的刹车能够增加车闸和轮胎之间的摩擦，从而让轮胎停止转动。不同材料会产生大小不同的摩擦力，通常情况下，光滑的表面上产生的摩擦力较小，粗糙的表面上产生的摩擦力较大。

有时我们需要减少摩擦力。例如，人们会用油来减少汽车引擎运动部分之间的摩擦。油能够润滑物体各部分的表面，使其顺畅运行。

刹车是利用摩擦力来使车轮停止转动的。

溜冰者穿着能产生较小摩擦力的特殊冰鞋。

3 请列出利用摩擦力来防止人们滑倒的各种鞋子。
4 哪种运动需要产生很小的摩擦力的装备？

Air Resistance

Air resistance is a force and is another type of friction. Just as two surfaces sliding against each other produce friction, when an object travels through the air there is friction between the air and the object. As the air moves over the surface of the object it produces friction, which we call air resistance.

Word Box
air resistance
streamlined

1a Have you ever noticed air resistance slowing you down?

1b What were you doing?

1c How did you reduce the effects?

The shape of an object can reduce the amount of air resistance. The air moves over the surface of a smooth curved object more easily than over a flat square-shaped object. We call shapes that move through air or water easily **streamlined**.

Vehicle designers try to reduce the effects of air resistance so that cars and lorries can travel more quickly and use less fuel.

2 Which vehicle moves more easily through the air?

空气阻力

空气阻力是一种力，同时也是另一种类型的摩擦力。正如两个物体的表面产生相对运动时会产生摩擦力，当一个物体在空气中运动时，物体和空气之间也会产生摩擦力。当空气在物体表面流过时，会产生摩擦力，我们将这种摩擦力叫作空气阻力。

单词表
air resistance 空气阻力
streamlined 流线型

1a 你是否注意到空气阻力可以使你的速度慢下来？
1b 你当时在做什么？
1c 你是如何降低空气阻力的影响的？

一个物体的形状能够减少空气阻力的大小。空气更容易通过呈曲线型的光滑物体的表面，而更难通过方形的平整物体的表面。我们将水或空气更容易通过的物体的形状称为流线型。

汽车设计商尝试降低空气阻力的影响，这样轿车和卡车的行驶速度更快且消耗燃料更少。

2 在空气中，哪种车行驶起来更容易？

Look at the photos above. They show just a few examples of how people have tried to reduce air resistance. Make a list of some other ways people have tried to reduce air resistance.

Sometimes we need to try and maximise air resistance. Skydivers use parachutes to maximise the effects of air resistance to slow them down.

3 What other forces act on a skydiver?

The surface area of the parachute increases air resistance and slows down the skydiver as they fall towards Earth.

观察上面的图片。这些图片展示了人们通过不同的方式减少空气阻力。请列出人们试图减少空气阻力的其他例子。

有时，我们也需要利用空气阻力或将空气阻力的效果最大化。跳伞运动员身上的降落伞可以将空气阻力的效果最大化，使其缓慢降落。

3 还有哪些力作用于跳伞运动员？

降落伞的表面区域能增大空气阻力，使跳伞运动员落到地面上的速度减少。

Science Skills

Paper Helicopters – Investigate it!

Word Box
variable

Ajay and Rhani investigated how paper helicopters fall through the air. They changed one **variable**, the length of the wings, each time and kept all other variables the same.
They dropped each helicopter from the same height and timed how long it took to reach the ground for different wing lengths. They repeated each measurement three times and recorded the results. They then worked out the mean average for the time taken for each helicopter to fall to the ground.

Here are the results:

Length of wings	Time #1 (seconds)	Time #2 (seconds)	Time #3 (seconds)	Mean average time (seconds)
7.5 cm	2.2	2.6	2.5	2.5
6.5 cm	2.1	2.0	2.2	2.1
5.5 cm	1.7	2.1	1.8	1.9
4.5 cm	1.7	1.6	1.7	1.6

科学技能

纸飞机——探究一下吧！

单词表
variable 变量

阿杰伊和拉尼对纸飞机如何在空中降落进行了探究。他们每次都改变同一个**变量**，即机翼的长度，同时保持其他变量一致。

他们让不同机翼长度的纸飞机从同一高度落下，并记录了每架纸飞机花了多长时间降落到地面。他们重复了三次实验并记录了结果，然后计算每架纸飞机落到地面所需时间的平均值。

结果如下：

机翼的长度	第一次落地所需时间（秒）	第二次落地所需时间（秒）	第三次落地所需时间（秒）	时间的平均值（秒）
7.5厘米	2.2	2.6	2.5	2.5
6.5厘米	2.1	2.0	2.2	2.1
5.5厘米	1.7	2.1	1.8	1.9
4.5厘米	1.7	1.6	1.7	1.6

> Use the results from the table to draw a graph. Use the graph and table of results to help you decide whether the conclusions that Rhani and Ajay have made below are accurate.

Rhani and Ajay's conclusions:
- Changing the wing length didn't make much difference to how quickly the helicopters fell.
- The shortest-winged helicopter fell the quickest.
- The longest-winged helicopter took the longest to fall to the ground.
- If the wings were 8.5 cm long the helicopter would take approximately 3 seconds to fall to the ground.
- If the wings were 3.5 cm long the helicopter would take 1 second to fall to the ground.

Making paper helicopters

1. Make your own paper helicopter.
2. What other variables could you change?

用表格中的结果绘制一张图表。根据图表和表格的结果来帮你判断下面阿杰伊和拉尼的结论是否准确。

阿杰伊和拉尼的结论：

- 改变机翼长度几乎不影响纸飞机下落的快慢。
- 机翼最短的纸飞机下落速度最快。
- 机翼最长的纸飞机下落时间最长。
- 如果纸飞机的机翼长度为8.5厘米，大约需要3秒落到地面上。
- 如果纸飞机的机翼长度为3.5厘米，大约需要1秒落到地面上。

制作纸飞机

1. 自己动手制作一个纸飞机。
2. 你还可以改变其他的变量吗？

Making Model Boats

You will need:

Word Box
hull

Large bowl of water

Swimming noodle float

Scissors

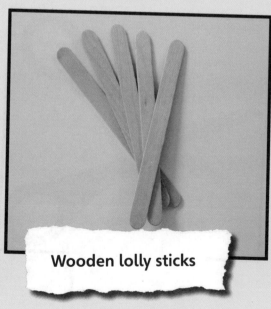
Wooden lolly sticks

制作模型船

你将需要：

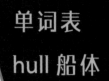

单词表
hull 船体

一大碗水

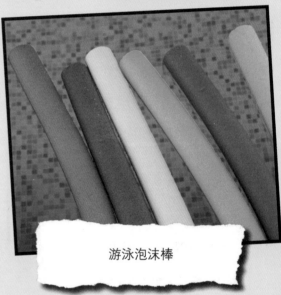

游泳泡沫棒

剪刀

木棍

Method

1. Cut the foam into different shapes to create a range of **hulls** (bases) for your boat.
2. Decide how to position the base of your boat, the hull, in the water. For example, the curved side on the bottom or flat side on the bottom.
3. Carefully cut the end of the wooden lolly stick to make a point.
4. Find the centre of the hull and insert the wooden mast (lolly stick). If the boat capsizes you may need to reposition your mast or turn the hull the opposite way round.

Make a small sail and then see whether your boat remains balanced in the water. Blow gently on the sail to move your boat.

方法

1. 将游泳泡沫棒裁剪成不同的形状，并将其作为**船体**（船的主体）。

2. 决定船体应该如何被放到水中。例如，是弯曲的一面朝下还是平坦的一面朝下。

3. 小心地将木棍的一端削尖。

4. 找到船体的中心，将木质桅杆（木棍）插入船体。如果船翻了，你就需要重新在别的位置插入桅杆，或者把船体的正反面翻转一下。

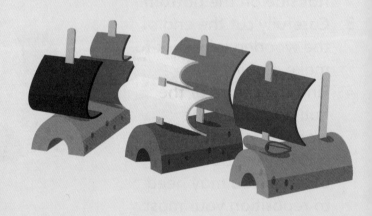

📖 制作一张小船帆，观察你的船是否能在水中保持平衡。轻轻地对着船帆吹气，让船移动。

Water Resistance

Water resistance is another force that slows things down. It is similar to air resistance and is a type of friction. As an object moves through water and the water passes over the surface of the object it causes friction and slows down the object.

Word Box
water resistance

The amount of water resistance can be reduced by using smooth surfaces. The shape of the object can be changed to make it more streamlined, for example, helping a boat to move through the water and move more quickly.

Boats have a thin pointed hull to reduce water resistance and move through the water faster.

The hull has a streamlined design to reduce water resistance.

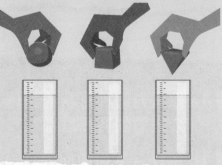

All three objects are dropped at the same time.

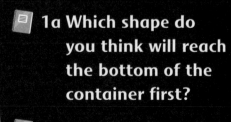

📖 **1a** Which shape do you think will reach the bottom of the container first?

📖 **1b** Explain your answer.

水阻力

水阻力是另外一种能减少物体运动速度的力。水阻力与空气阻力相似，也是一种摩擦力。当物体在水中移动时，水从物体表面流过会产生摩擦力，使物体的运动速度减少。

光滑的表面能使水阻力变小。可以把物体的形状改造得更偏流线型，例如，流线型的船就能在水中移动，且速度更快。

船有着薄薄的船体和尖尖的船头，这样可以减少水阻力，更快地在水中移动。

单词表
water resistance 水阻力

船体呈流线型设计，这样可以减少水阻力。

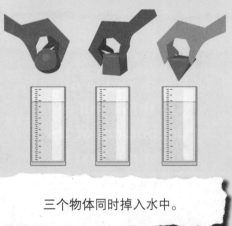

三个物体同时掉入水中。

- **1a** 你认为哪种形状的物体会先落到容器底部？
- **1b** 解释一下你的回答。

Simple Mechanisms

Forces can make things move, speed up, slow down or change direction. Forces are useful for many things in everyday life and we use simple machines, or **mechanisms**, to help us use force and make our lives easier. We can use machines so that we need less force to do the same job, and we can also use them to change the direction of a force. Here we look at three simple mechanisms: **gears**, **levers**, and **pulleys**.

Word Box
axle
gear
gearwheel
lever
mechanism
pulley

Gears

A **gearwheel**, or cog, has teeth around its edges that interlock with similar teeth on another wheel. As one wheel turns it causes the other to turn at the same time. Using interlocking gears of different sizes can make it much easier to move things.

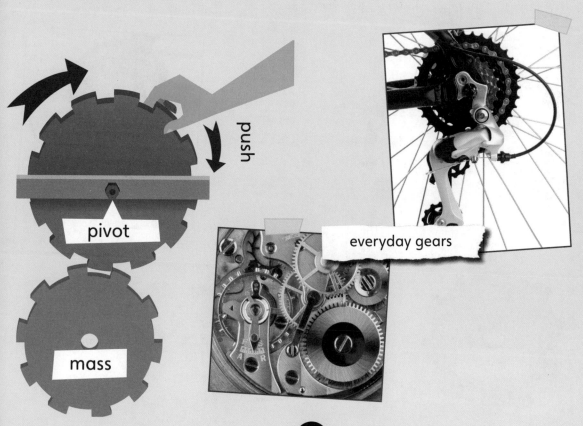

pivot

push

mass

everyday gears

简单的机械装置

力能让物体移动、加速、减速或改变运动方向。力在我们日常生活中有很多作用，我们会使用一些简单的机械或**机械装置**，来帮助我们借助力的作用，让我们的生活变得更轻松。通过使用机械，我们可以运用更少的力来完成同样的工作，同时，我们也能利用机械来改变力的方向。我们来认识一下三种简单的机械装置：**齿轮**、**杠杆**和**滑轮**。

单词表
axle 轴
gear 齿轮装置；齿轮
gearwheel 齿轮
lever 杠杆装置；杠杆
mechanism 机械装置
pulley 滑轮装置；滑轮

齿轮装置

齿轮是边缘为锯齿状的轮子，利用的是轮齿的互相啮合。当一个齿轮转动时会带动另一个齿轮同时转动。利用大小不同的、啮合在一起的齿轮，我们移动物体就变得更容易了。

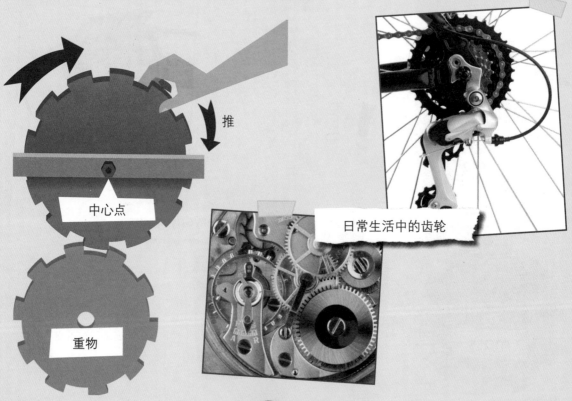

推

中心点

重物

日常生活中的齿轮

Levers

Have you noticed how much easier it is to lift someone when they are sitting on a see-saw? It is easier because the see-saw is a lever. A lever is an arm that turns, or pivots, around a point, and can turn a small force into a larger one.

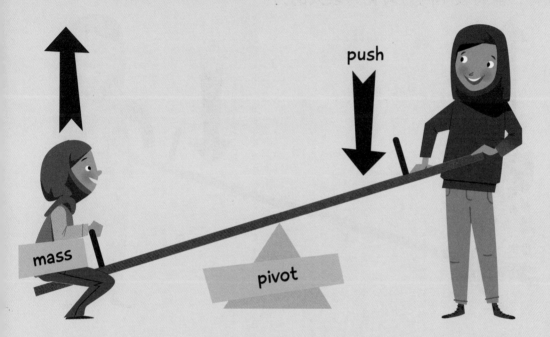

Levers are useful in lots of ways, for example, orangutans have learned to use levers to help them open fruit!

Everyday levers

杠杆装置

你是否注意到，当一个人坐在跷跷板上时，我们可以更轻易地抬起他/她？这是因为跷跷板是一个杠杆。杠杆是一种可以绕着支点旋转的长杆，能将较小的力转化为较大的力。

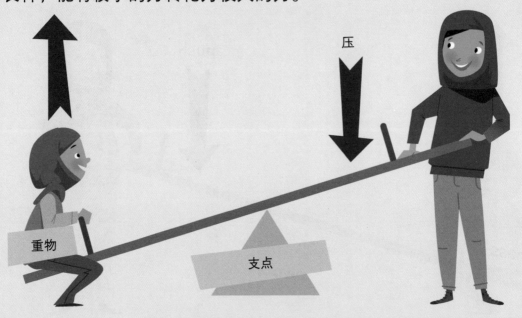

杠杆在很多方面都是有用的，例如，猩猩能利用杠杆帮自己撬开水果！

日常生活中的杠杆

Pulleys

A pulley has a grooved wheel which spins on a rod called an **axle**. A rope is threaded around the wheel and attached to an object. Pulleys are very useful mechanisms as they help us use less force to lift heavy objects.

Everyday pulleys

A force of 10 N is needed to lift a 1 kg mass. By using a pulley the force needed to lift the mass is spread between the different lengths of rope. This makes it is easier to pull up the mass.

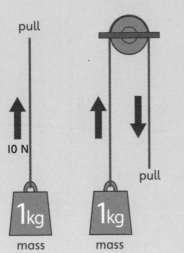

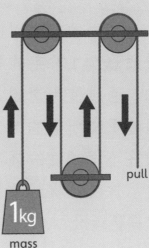

滑轮装置

滑轮是一种周围带凹槽且可以绕**轴**旋转的圆轮。绳索可以跨过圆轮上的凹槽连接物体。滑轮是十分有用的机械装置,因为它能帮助我们用更少的力提起重物。

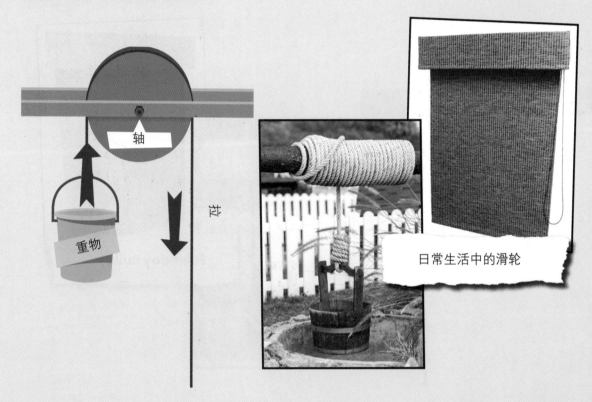

日常生活中的滑轮

10牛的力可以提起质量为1千克的物体。通过使用滑轮,提起重物所需的力分散在不同长度的绳索上。这就让提起重物变得更轻松。

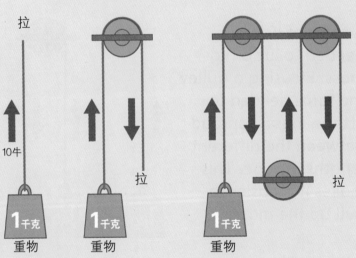

Forces

Use the words below to create a concept map to show what you already know about forces. You can add your own words to the list.

force   surfaces   push   pull

What Forces Do

Make a list of sports that could be in a sports competition in the table below. Write how each sport uses forces to make things move.

Sport	Forces used to move things

Sir Isaac Newton

Research the life and work of Sir Isaac Newton. You could present your findings as a mini-book. Use the space below to plan the pages for your mini-book.

Newton Meters

Use a newton meter to find what force is needed to lift some everyday objects.

Object	Force (N)
School bag	

Life with Less Gravity

What would life be like if there were *less* gravity on Earth? Use the space below to record your ideas. You can use words and pictures to complete your work.

Life with More Gravity

What would life be like if there were *more* gravity on Earth? Use the space below to record your ideas. You can use words and pictures to complete your work.

What Have You Learned So Far?

Think about everything you have learned so far. Write some questions to test your classmate's knowledge of forces. The first question has been written for you. Don't forget to make a note of the answers.

What is gravity?

Gravity is a pull force that makes things fall to the ground and stay there.

Science Skills

Gravity on Other Planets – Graph it!

1. Draw a graph to show how heavy a bag of sugar would feel on different planets. Use the information in the table below.
2. Label the planets that have a similar gravity to Earth.

Planet	Mass of sugar	Relation to Earth's gravitational pull (approx.)	Weight
Mercury	1 kg	1/3rd	3.8 N
Venus	1 kg	9/10th	9.1 N
Earth	1 kg		10 N
Mars	1 kg	1/3rd	3.8 N
Jupiter	1 kg	2.5 times stronger	23.6 N
Saturn	1 kg	Slightly stronger	10.6 N
Uranus	1 kg	4/5th	8.9 N
Neptune	1 kg	1/10th stronger	11.3 N

Exploring Friction

Look at the diagram below.
1. Draw a circle around the areas where friction might be produced.
2. Add speech bubbles to explain your answers.

Investigating Shoes

1. Use the space below to draw a picture of the sole of your shoe. Look carefully for any patterns you can see on the sole.

2. What do you notice about the sole of your shoe? Why do you think there are patterns?

Grip or Slip?

Produce an advert to recommend the best shoes for slipping. Here's an example:

if you want to slip and slide then these are the shoes for you!

Then produce an advert for shoes that have great grip. Your advert should have a slogan and a labelled picture that shows why the shoes are so grippy.

NEW!

The sole is smooth!

No ridges here!
The smooth surface reduces friction so slipping is easy.

CAUTION SLIPPY

Exploring Air Resistance

Think about when you travel outside, you may be running, walking, on your bike or on a skateboard.

1. What can you see or feel that tells you there is air resistance when you move? For example, clothes move in the wind.

   _____
   _____

2. Do you feel more or less air resistance when you move quickly? How can you tell?

   _____
   _____

3. Use the space below to draw a labelled diagram to explain how air resistance works. Think about it in terms of forces.

Falling Objects

Lots of objects are designed to use air resistance to change how they move. Parachutes, boat sails, shuttlecocks and kites all use air resistance to change how they move.
Look at the objects below.

1 What do you think would happen if they were thrown into the air?

2 How would air resistance affect how they fall to the ground?

3 Record your observations in the space below using labelled diagrams.

Science Skills

Paper Helicopters – Investigate it!

Ajay and Rhani were investigating how paper helicopters fall. Here are the results:

Length of wings	Time #1 (seconds)	Time #2 (seconds)	Time #3 (seconds)	Mean average time (seconds)
7.5 cm	2.2	2.6	2.5	2.5
6.5 cm	2.1	2.0	2.2	2.1
5.5 cm	1.7	2.1	1.8	1.9
4.5 cm	1.7	1.6	1.7	1.6

1 Use the results to draw a graph.

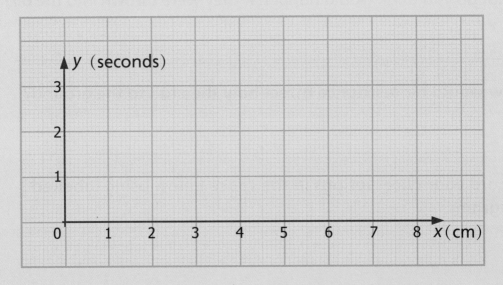

2 Draw a line through the inaccurate conclusions.

Rhani and Ajay's conclusions:
- Changing the wing length didn't make much difference to how quickly the helicopters fell.
- The shortest-winged helicopter fell the quickest.
- The longest-winged helicopter took the longest to fall to the ground.
- If the wings were 8.5 cm long the helicopter would take approximately 3 seconds to fall to the ground.
- If the wings were 3.5 cm long the helicopter would take 1 second to fall to the ground.

Now make your own paper helicopter.

3 Use what you have learned from your investigation of paper helicopters to write some instructions explaining how to make the best paper helicopter.
4 Remember to explain why your helicopter is the best.
   Does it spin a lot, stay in the air for a long time or change direction as it falls to the ground?

5 What variables could be changed on your helicopter?

Making Model Boats

Sail boats use the wind to overcome the effects of air resistance and water resistance to move a boat through the water. The wind pushes against the sails, if the wind is stronger than the air and water resistance the boat will move forward.

1a Which sail helped the boat move the quickest? Use the space below to draw around the sail and record your answer.

1b Explain why this sail worked the best.

2   How could you improve the design?

Water Resistance

Water resistance is a type of friction that slows down things that are moving in water.

1. How does changing the shape of the modelling clay change how it moves in water?

2a Which shape do you think will reach the bottom of the container first?

2b Which shape do you think will fall the slowest?

2c Explain your answers.

3. Which other shapes do you think would be good at reducing the effects of water resistance?

Designing a Fantastic Boat

1. Use everything you have learned about the different types of friction to design an amazing boat.
   Your boat must be able to overcome the effects of air resistance and water resistance.
2. Label your diagram to show all the places where friction might occur.

Forces Are All around Us

1 Fill in the missing words. There is a list of words at the bottom to help you.

Forces affect every aspect of our lives. They can make things _____ or they can _____ things down.

They can even stop them from moving completely.

_____ happens where there is contact between two moving surfaces. Other _____ can act at a distance without touching, like _____ and magnetism.

Air resistance and water resistance are both types of_____. If the surface of the object moving through the water or air is _____ there will be less friction. Cars, planes, trains and boats all try to reduce the amount of friction produced by creating _____ shapes.

This helps the air and the water to easily move over the vehicle which makes _____ less difficult.

forces   friction   gravity   move   moving
slow   smooth   streamlined

2 Create your own forces sentence with missing words. Show it to a classmate and ask them to fill in the gaps.

Grouping Mechanisms

Think about the simple machines we use in our everyday lives.

Use the space provided to group your ideas using the headings provided. You can use drawings or lists to complete your work.

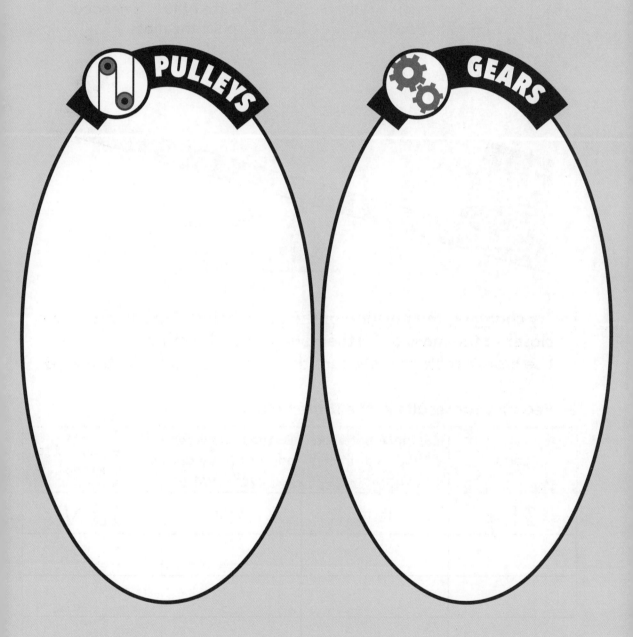

Levers

Levers are a simple mechanism that can make a job easier to do.

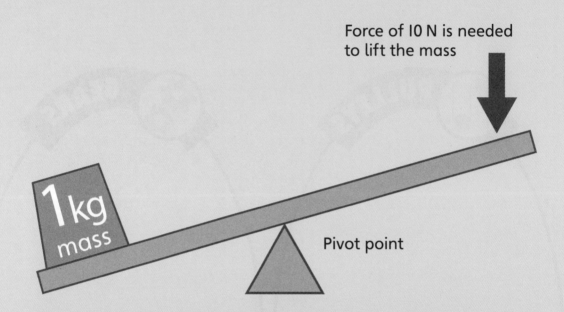

1. Try changing the position of the pivot point. You can move it closer to the mass or further away from the mass.
   Use a push meter to measure the force needed to lift the load.

2. Record your results in the table below.

Load	Distance between load and pivot point	Distance between point of force and pivot point	Force needed to lift load
1 kg	15 cm	15 cm	10 N

Fun with Forces

1. Imagine you are a superhero with the power to increase the effect of one force. What force would you choose and why?

2. What would your superhero name be?

3. Use the space below to create a cartoon strip to show how you would use your superhero powers.

What I Know about Forces

Think about what you have learned in this unit.

1 **What have you learned?**

2 **What did you find interesting?**

3 **What would you like to research in more detail?**

Glossary

air resistance: the frictional force air exerts against a moving object

axle: a rod that fits through the centre of a wheel

Earth: the planet on which we live

force: causes a change in motion of an object, so that the object moves or slows down or speeds up, or changes direction

friction: a force between two surfaces that are sliding, or trying to slide, across each other

gear: simple machine that uses interlocking cogged wheels

gearwheel: a wheel with teeth

gravitational pull: the amount of force of gravity an object exerts

gravity: force that causes unsupported objects to fall

hull: the part of the boat that is submerged in water

lever: simple machine that moves on a pivot point

mass: a measure of how much matter is in an object, with the unit of grams or kilograms

mechanism: devices designed to make it easier to move objects

move: change position

newton (N): unit of measure for force

newton meter: equipment used to measure force

orbit: the path planets take around the Sun

pull: type of force used to make things move

pulley: simple machine that has a grooved wheel that spins on an axle

push: type of force used to make things move

space: the empty area outside Earth's atmosphere, where the planets and the stars are

streamlined: when an object has been shaped to reduce friction in air or water

tide: the rise and fall of sea levels caused by the Sun and Moon's gravitational pull

variable: anything in an investigation that can be controlled or changed

water resistance: type of friction that slows down objects moving in water

weight: the force exerted on a mass due to gravity

词汇表

空气阻力：空气对运动中的物体施加的摩擦力
轴：固定在圆轮中间的长杆
地球：人类生活的行星
力：能改变物体的运动状态，使其发生移动、减小或增大运动速度，以及改变运动方向
摩擦力：当一个物体在另一个物体的表面产生相对位移或将要产生相对位移时，在两个物体的接触面上产生的作用力
齿轮装置：一种由相互啮合的齿轮组成的简单机械装置
齿轮：边缘为齿状的圆轮
万有引力：一个物体所施加的引力的大小
地心引力：使没有支撑的物体下落的力
船体：没入水中的船的主体部分
杠杆：绕着支点运动的简单机械装置
质量：用来衡量一个物体所含物质的多少，以克或千克为单位
机械装置：人们设计出的能够更轻易地移动物体的装置
移动：产生位移
牛顿（N）：力的单位
弹簧测力计：测量力的大小的设备
轨道：行星绕太阳运行的路径
拉力：一种让物体移动的力
滑轮：周围带凹槽且可以绕轴旋转的圆轮
推力：一种让物体移动的力
太空：地球大气层以外的区域，太空中有行星和恒星
流线型：可减少在空气中或水中的摩擦力的物体的形状
潮汐：由月球和太阳的万有引力作用造成的海洋水平面定时涨落的现象
变量：研究中可以人为控制或改变的因素
水阻力：让物体在水中的移动速度减小的一种摩擦力
重力：物体所受的地心引力

培生科学虫双语百科

奇妙物理

Light and Sight

光与视觉

英国培生教育出版集团 著·绘

徐 昂 译

电子工业出版社

Publishing House of Electronics Industry

北京·BEIJING

Original edition, entitled SCIENCE BUG and the title Light and Sight Topic Book, by Tara Lievesley published by Pearson Education Limited © Pearson Education Limited 2018
ISBN: 9780435196554

All rights reserved. No part of this book may be reproduced or transmitted in any form or by any means, electronic or mechanical, including photocopying, recording or by any information storage retrieval system, without permission from Pearson Education Limited.

This adaptation of SCIENCE BUG is published by arrangement with Pearson Education Limited. Chinese Simplified Characters and English language (Bi-lingual form) edition published by PUBLISHING HOUSE OF ELECTRONICS INDUSTRY, Copyright © 2023.

For sale and distribution in the mainland of China exclusively (except Hong Kong SAR, Macau SAR and Taiwan).

本书中英双语版由Pearson Education（培生教育出版集团）授权电子工业出版社在中华人民共和国境内（不包括香港、澳门特别行政区及台湾地区）独家出版发行。未经出版者书面许可，不得以任何方式抄袭、复制或节录本书中的任何部分。

本套书封底贴有Pearson Education（培生教育出版集团）激光防伪标签，无标签者不得销售。

版权贸易合同登记号　图字：01-2022-2381

图书在版编目（CIP）数据

培生科学虫双语百科. 奇妙物理. 光与视觉：英汉对照／英国培生教育出版集团著、绘；徐昂译. --北京：电子工业出版社，2024.1
ISBN 978-7-121-45132-4

Ⅰ.①培… Ⅱ.①英… ②徐… Ⅲ.①科学知识－少儿读物－英、汉 ②光学－少儿读物－英、汉 Ⅳ.①Z228.1 ②O43-49

中国国家版本馆CIP数据核字（2023）第035082号

责任编辑：李黎明　文字编辑：王佳宇
印　　刷：河北迅捷佳彩印刷有限公司
装　　订：河北迅捷佳彩印刷有限公司
出版发行：电子工业出版社
　　　　　北京市海淀区万寿路173信箱　邮编：100036
开　　本：787×1092　1/16　印张：35　字数：840千字
版　　次：2024年1月第1版
印　　次：2024年2月第2次印刷
定　　价：199.00元（全9册）

凡所购买电子工业出版社图书有缺损问题，请向购买书店调换。若书店售缺，请与本社发行部联系，联系及邮购电话：（010）88254888，88258888。
质量投诉请发邮件至zlts@phei.com.cn，盗版侵权举报请发邮件至dbqq@phei.com.cn。
本书咨询联系方式：010-88254417，lilm@phei.com.cn。

使用说明

欢迎来到少年智双语馆！《培生科学虫双语百科》是一套知识全面、妙趣横生的儿童科普丛书，由英国培生教育出版集团组织英国中小学科学教师和教研专家团队编写，根据英国国家课程标准精心设计，可准确对标国内义务教育科学课程标准（2022年版）。丛书涉及物理、化学、生物、地理等学科，主要面向小学1~6年级，能够点燃孩子对科学知识和大千世界的好奇心，激发孩子丰富的想象力。

本书主要内容是小学阶段孩子需要掌握的物理知识，含9个分册，每个分册围绕一个主题进行讲解和练习。每个分册分为三章。第一章是"科学虫趣味课堂"，这一章将为孩子介绍科学知识，培养科学技能，不仅包含单词表、问题和反思模块，还收录了多种有趣、易操作的科学实验和动手活动，有利于培养孩子的科学思维。第二章是"科学虫大闯关"，这一章是根据第一章的知识点设置的学习任务和拓展练习，能够帮助孩子及时巩固知识点，准确评估自己对知识的掌握程度。第三章是"科学词汇加油站"，这一章将全书涉及的重点科学词汇进行了梳理和总结，方便孩子理解和记忆科学词汇。

2024年，《培生科学虫双语百科》系列双语版由我社首次引进出版。为了帮助青少年读者进行高效的独立阅读，并方便家长进行阅读指导或亲子共读，我们为本书设置了以下内容。

（1）每个分册第一部分的英语原文（奇数页）后均配有对应的译文（偶数页），跨页部分除外。读者既可以进行汉英对照阅读，也可以进行单语种独立阅读。问题前面的 📖 符号表示该问题可在第二部分预留的位置作答。

（2）每个分册第二部分的电子版译文可在目录页扫码获取。

（3）本书还配有英音朗读音频和科学活动双语视频，也可在目录页扫码获取。

最后，祝愿每位读者都能够享受双语阅读，在汲取科学知识的同时，看见更大的世界，成为更好的自己！

电子工业出版社青少年教育分社
2024年1月

Contents 目录

Part 1　科学虫趣味课堂　　　　　　　　　　　　　　／1

Part 2　科学虫大闯关　　　　　　　　　　　　　　　／49

Part 3　科学词汇加油站　　　　　　　　　　　　　　／73

Part2译文

配套音视频

Light and Sight

You cannot hear it, taste it, feel it or smell it... but you can see it. What is it? Light! We need our eyes to see objects around us. We also need a light source. Some people say that they can see at night time, when it is dark. Do you think this is true?

1 Why can we often see at night time?
2 What would you see if there was a complete absence of light?

What do you know about light?

Discuss with your group what you already know about light and what questions you could ask about light.

How do you think we see?

1 What two things do we need to be able to see an object?
2 How do you think these two things work together so you can see something? Draw your ideas and describe your image to your group.

光与视觉

你听不到它、尝不到它、感觉不到它、闻不到它……但是你能看到它。它是什么呢？是光！我们需要通过眼睛来观察我们周围的物体，我们同样也需要光源。一些人说他们能在夜晚的黑暗中看清物体，你认为这是真的吗？

1 为什么我们在夜晚能看见物体？
2 如果完全没有光，你能看见什么？

关于光，你知道什么？

和你的小伙伴讨论一下你所知道的关于光的知识，以及你们想问的关于光的问题。

你认为我们是如何看见物体的？

1 我们需要具备哪两个条件才能看见物体？
2 你认为这两个条件是如何发挥作用让你看见物体的？将你的想法画出来并向你的小伙伴描述一下。

All about Light Sources

Some sources of light are natural, such as the Sun, which is our most important light source. Some light sources are **artificial**, or human-made, such as the headlights on a car.

Word Box
artificial
reflect

How many light sources?

Think of as many different light sources as you can. Make two lists, one of natural light sources and one of human-made light sources. Which list is longer? Why do you think this is?

1 Discuss in your group how our lives would be different if we did not have artificial light sources.

All light sources help us to see.

Before electric lights were invented, people moulded wax into candles and lit them for light. The wax has a piece of cord or string in the centre to act as a wick. Wax can be made from bees, the fat from animals, or even from oil. The wick could be made from animal hair or from plant fibres.

2a Do you think a candle is a natural or a human-made source of light?

2b Discuss points for both ideas.

2c Explain your final decision to the rest of your group.

关于光源的一切

一些光源是自然光源，例如，我们最重要的光源——太阳。一些光源是**人造**光源，例如，汽车的前灯。

单词表
artificial 人造的
reflect 反射

有多少种光源？

请尽可能地思考一下有哪些不同的光源。将你想到的光源列成表格并分为两种，一种是自然光源，另一种是人造光源。哪张表格更长？为什么？

1 和你的小伙伴一起，讨论一下，如果我们没有人造光源，我们的生活会有哪些不同。

所有的光源有助于我们看见物体。

在电灯被发明以前，人们将蜡制成蜡烛，点燃它们可以获得光。蜡烛中间有一根细绳作为灯芯，蜡可以由蜜蜂、动物脂肪甚至油制成。灯芯是由动物毛发或植物纤维制成的。

2a 你认为蜡烛是自然光源还是人造光源？

2b 讨论一下两种观点。

2c 和你的小伙伴解释一下你的最终看法。

Not all things that are bright, shiny and can be seen easily are sources of light. Sometimes we can confuse a source of light with something that is reflecting light. For example, the Sun is a source of light, but the Moon is not, because the Moon **reflects** the light from the Sun. It is dangerous to look directly at the Sun, but our eyes will not be harmed if we look directly at the Moon.

Some of the objects shown on these pages are human-made and some are natural.

3 Use a Carroll diagram like the one below, to sort the objects into four groups. Do you notice anything interesting about the groups?

	Human-made	Natural
Light source		
Reflector		

A Carroll diagram is used to group and sort things using two different characteristics.

4 How could you define a light source, so that it is not confused with a reflector of light?

5 Use this definition to help explain why a mirror is not a light source.

不是所有明亮的且能轻易被看见的物体都是光源。有时,我们会将光源和反射光的物体混淆。例如,太阳是光源而月亮不是,因为月亮**反射**了来自太阳的光。我们直视太阳是很危险的,但如果我们直视月亮,我们的眼睛并不会受到伤害。

书中呈现的这些物体有些是自然光源,有些是人造光源。

3 请将下面的物体分为四类,填进下面的卡罗尔图表中。关于分组,你注意到哪些有趣的现象?

	人造的	自然的
光源		
反光体		

卡罗尔图表可以通过两种不同的特性将物体进行分类。

4 如何定义光源,才不会把它与反光体混淆?

5 请借助你的定义,解释一下为什么镜子不是光源。

Travelling Light

We do not see objects instantly as light takes time to travel to our eyes. As light can travel 300,000 kilometres every second (**light speed**) this happens very fast. Light is even faster than sound; nearly 900,000 times faster! Nothing can travel faster than light.

Word Box
light speed

Light speed

Every second, light can travel seven times around the Earth. Draw a circle on a piece of paper. This represents the Earth. Ask a friend to time you as you draw around the circle seven times, as fast as you can. How long did it take? Can you draw at the speed of light? Why?

Jet aeroplanes can travel very fast. This one travels at 2 km per second (7,200 km per hour) and takes 5.5 hours to travel around the Earth.

This star is the closest star to Earth, after the Sun. The light it produces takes 4.3 light years to reach us on Earth.

You will need a calculator to help you explore these questions.
1. The Sun is 148 million kilometres from the Earth. Calculate how long it takes light from the Sun to reach Earth.
2. Calculate the circumference of the Earth, using light speed and how long it takes to travel around the Earth seven times.
3. A light year is the distance light can travel in a single year. Calculate how far this is, in kilometres.

传播的光

我们不能立刻看见物体，这是因为光到达我们的眼睛需要时间。光传播得非常快，以300,000千米每秒的速度传播（**光速**）。光的传播速度比声音的传播速度快很多，几乎是90万倍！没有什么物体的速度比光速更快了。

单词表
light speed 光速

光速

每隔一秒，光就能绕着地球传播七周。在纸上画一个圆，这个圆代表地球。再绕着圆画七次，越快越好，让你的朋友为你计时，你花费了多长时间？你画画的速度能达到光速吗？为什么？

喷气式飞机可以飞行得非常快。这架飞机的速度能达到2千米每秒（7,200千米每小时），它只要5.5小时就能绕着地球飞行一周。

这颗恒星是除太阳外离地球最近的恒星，它产生的光需要传播4.3光年才能到达地球。

你将需要一个计算器来帮你解答这些问题。
1. 太阳与地球的距离是148,000,000千米，请计算光从太阳到达地球所需的时间。
2. 请结合光速和光绕地球传播七周所需的时间，计算地球的周长。
3. 一光年是光在一年内传播的距离，请计算一光年等于多少千米？

Light Rays

In order for light to travel so fast it travels in straight lines, which scientists call **rays**. Light rays do not bend, but they can change direction when they hit something smooth, flat and shiny.

Here, straight rays of light are shining through the leaves in a forest.

Word Box
ray

You will need: a torch, a long cardboard tube, three pieces of card, a black piece of paper, a comb...

We can demonstrate that light travels in straight lines with some different experiments. Try the following ideas:

1 Look at a torch beam through a long cardboard tube. Now bend the tube, point one end at the torch and look down the other end. What can you see? Why?

2 Make a hole in the centre of three pieces of card. Place the cards in a line, with a black piece of paper at one end of the line and a torch at the other end. Shine a torch at the holes. What can you see? Why? Now lift up one of the pieces of card, so the holes are not lined up and shine the torch. What happens? Why?

3 Shine a torch through a comb. What can you see? Why?

1 How do each of the experiments prove that light travels in straight lines?

2 Write a short news report to explain how light travels. Include some diagrams to help explain.

3 Discuss in your group what would be different about this diagram if light rays travelled in a curve or a zigzag.

光线

为了传播得更快,光是沿直线传播的,科学家们称之为**光线**。光线不能弯曲,但是当光线接触到光滑、平整、明亮的表面时,可以改变方向。

在森林里,笔直的光线透过树叶照射下来。

单词表
ray 光线

> 你将需要:一只手电筒、一根硬纸管、三张卡片、一张黑纸、一把梳子……
>
> 我们可以通过不同的实验来证明光是沿直线传播的。尝试下面的想法:
>
> 1 透过硬纸管观察一束手电筒发出的光。然后将硬纸管弯曲,再将手电筒对着纸管的一端打开,从另一端看过去,你能观察到什么?为什么?
>
> 2 在三张卡片的中心剪出一个洞。将卡片排成一条直线,把黑纸放在三张卡片的一边,手电筒放在三张卡片的另一边。对着洞打开手电筒。你能观察到什么?为什么?再试着把其中一张卡片抬高,这样三个洞就不在一条直线上了,然后再打开手电筒。现在你能观察到什么?为什么?
>
> 3 对着梳子打开手电筒。你观察到了什么?为什么?

1 每个实验是如何证明光是沿直线传播的?

2 请写出一个简短的报告来解释光是如何传播的。可以运用一些图表来解释。

3 和你的小伙伴一起,讨论一下,如果光线是沿曲线传播的或是沿"之"字形传播的,这张图会有什么不同?

Light for Seeing

Light rays **radiate** out at very high speed, in straight lines from a light source and enter our eyes. The light enters through the black hole in the centre of our eye, called the **pupil**. But how do we see objects other than light sources?

Word Box
pupil
radiate

Thought experiment

Imagine putting a black piece of paper into a black box then putting the lid on the box. If you could see inside the box, what would you see? Why? Would it make any difference if the piece of paper was white instead of black? Why?

Seeing things

You will need: a piece of black paper, a cardboard tube, a coin or paperclip...

Roll up a piece of black paper into a tube. Place the tube over a small object, such as a coin or a paperclip. Look down the tube.

1 What can you see when you look down the tube? Why?

2 What happens if you make a hole in the side of the tube and shine a torch in?

3 Use the evidence from this activity to help explain and draw a picture of how we see objects in the tube.

光让我们能看见

光线以很快的速度从光源**发出**,沿直线传播,进入我们的眼睛。光是从我们眼睛中心的黑孔进入的,人们称黑孔为**瞳孔**。但是我们如何看到不是光源的其他物体呢?

单词表
pupil 瞳孔
radiate 辐射;发出

思想实验

想象一下,将一张黑纸放进一个黑箱子里,再盖上箱子的盖子。如果你能在箱子里观察,你会观察到什么?为什么?如果这张纸是白色的而不是黑色的,会有什么不同吗?为什么?

📓 看见物体

你将需要:一张黑纸、一根硬纸管、一枚硬币或回形针……

将一张黑纸卷起来放进硬纸管里。将硬纸管竖着放到一个小物体上,小物体可以是硬币或回形针。从硬纸管的上方向下观察。

1 当你向下观察时,观察到了什么?为什么?
2 如果你在硬纸管的一侧剪出一个洞,用手电筒照射洞口,会发生什么?
3 运用这个实验中发现的证据,来解释我们是如何看到硬纸管里的物体的,试着画张示意图。

Light Paths

We cannot see the path light takes to our eyes, but we can show on diagrams what we think is happening. We draw light rays with a ruler (as light travels in straight lines) and we place an arrow on the line to show which way the light is travelling. These are called **ray diagrams**.

Word Box
ray diagram

Light travels from the source and bounces from the object into our eyes. It is important to draw the arrow facing the correct direction.

1 Discuss what is wrong with this diagram.

2 If light really did travel in this direction what would we see when we looked at another person?

光路

我们不能看到传播到我们眼睛里的光走过的路径，但是我们能用示意图画出我们认为在发生的事情。我们可以借助直尺画出光线（因为光是沿直线传播的），然后在光线上标注箭头来表示光的传播方向。这样的图叫作**光路图**。

单词表
ray diagram 光路图

光从光源发出，然后在物体上反射，最终进入我们的眼睛。画出表示正确的传播方向的箭头是重要的。

1 讨论一下，下面这张示意图有什么错误。

2 如果光的传播方向真的如上图所示，那么当我们看向其他人时，我们能观察到什么？

How do we see?

Have you ever heard people talking about someone having 'piercing eyes' or 'throwing a glance'? Everyday language can make it seem as if our eyes are actively producing mysterious beams which let us see, but this is not true. Our eyes receive light that has been reflected off the objects around us, then our brain makes sense of what we see.

Did you know?

About 1000 years ago, scientist Al Hazan, knew we saw objects because of reflected light travelling to our eyes.

Things to do

Take a mirror into a totally dark place. Why can't you see anything in it?

我们是如何看见的？

你是否听到人们议论其他人有"犀利的眼神"或进行"瞥一眼"的动作？像这样的日常表达似乎表明了我们的眼睛在自主地产生神秘的光束，让我们可以看见物体，但事实并非如此。我们的眼睛接收到从周围物体反射过来的光线，然后大脑对我们看到的信息进行加工。

你知道吗？

一千多年前，一位名叫阿尔哈赞的科学家发现，我们之所以能看见物体，是因为被反射的光线进入了我们的眼睛。

试一试

将镜子拿到一个完全黑暗的地方。为什么从镜子里你什么也观察不到？

The Eye

We see an object when some of the reflected light enters our eyes. The light has travelled from the source to the object and then to our eyes.

Word Box
interpret
light-sensitive
optic nerve
transmit

An Arabian scientist called Alhazen (also known as Al-Hasan Ibn al-Haytham) was the first to explain how we see, over 1,000 years ago.

Find out more about Alhazen and his discoveries.

Once the light has entered our eyes, this is not the end of the journey. The light travels through our eye, and to the **light-sensitive** cells at the back of our eyes. These cells **transmit** a message to our brain, along the **optic nerve**, which **interprets** the message so we 'see' the object.

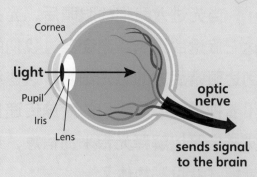

Pupil role

Cover your eyes with your hands. Face your partner. Count to ten. Quickly remove your hands. Ask your partner to watch your eyes, particularly your pupils. Discuss what your partner saw. Swap around and observe your partner's pupils. Record your observations.

1 Use your observations from the Pupil role activity to explain why your pupil behaves this way. How does this protect our eyes?

2 Find out what the coloured part of your eye is called.

3 Eyelids, tears and eyelashes also play an important role for our eyes. What is it?

眼睛

当物体反射的光线进入我们的眼睛时，我们就能看见这个物体了。光从光源射向物体，然后经反射到达我们的眼睛。

单词表

interpret 解释
light-sensitive 感光的
optic nerve 视觉神经
transmit 传递

一千多年前，名叫阿尔哈赞（阿尔–哈桑·伊本·阿尔–海瑟姆）的阿拉伯科学家成了第一位解释我们是如何看见物体的人。

找出更多关于阿尔哈赞的信息和他的发现。

当光进入我们的眼睛后，光并没有到达它旅途的终点。光会穿过我们的眼睛，到达眼睛后方的**感光**细胞。这些细胞可以沿着**视觉神经**向我们的大脑**传递**信息，视觉神经可以**解释**光带来的信息，这样我们就能"看见"物体了。

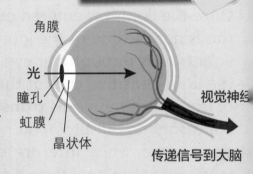

瞳孔的作用

用双手遮住你的眼睛。
面朝你的小伙伴。从一数到十。
快速地把手移开。
让你的小伙伴观察你的眼睛，特别是你的瞳孔。
和你的小伙伴讨论一下，他/她观察到了什么。
然后互换，由你来观察小伙伴的瞳孔。
记录一下你们的观察结果。

1. 结合你在瞳孔观察实验中的发现，解释一下为什么你的瞳孔会这样。怎样做能保护我们的眼睛？
2. 确定眼睛中带颜色的部分的名称。
3. 眼睑、眼泪和睫毛对于我们的眼睛很重要。它们分别有哪些作用？

How We See

When light travels from the source and reaches an object, it bounces off the object. Some objects look shiny, such as mirrors. These have very smooth surfaces and reflect most of the light in the same direction. Scientists call this bouncing of light off a surface, a **reflection**.

Word Box
reflection
scatter

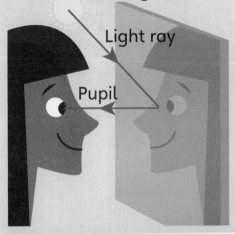

1 What do you think this diagram shows? Describe it to your group.

Some surfaces may appear smooth but are actually quite uneven. If the surface of the object is uneven, the light that bounces off the object is **scattered** in lots of different directions: the object will appear dull.

2 Which of these two surfaces would appear dull and which would be shiny? Explain your ideas to your group.

Light rays bouncing off a smooth (left) and uneven (right) surface.

You will need: a torch.

Are there more shiny or more dull objects in the room where you are?

1 Use a torch as your light source to shine light on objects in the room. Draw a table to record your observations.

2 Draw ray diagrams of the objects you shone the torch on to show how the light behaves when the light hits the object.

我们是如何看见的

当光从光源发出并到达某一物体时，会从物体表面反射。一些物体看上去很闪亮，例如，镜子。这些物体有十分光滑的表面，能够反射来自同一方向的大多数光线。科学家们把光从物体表面反弹的现象叫作**反射**。

单词表
reflection 反射；镜像
scatter 散射

1 你认为左图展示了什么现象？向你的小伙伴描述一下吧。

一些物体的表面看上去是光滑的，但实际上是不太平整的。如果物体表面不平整，那么这个物体反射的光线就会**散射**在不同方向，这导致物体看上去暗淡无光。

2 这两个表面哪一个看上去比较暗？哪一个比较亮？向你的小伙伴解释一下你的想法吧。

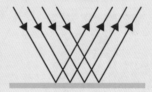 

光线分别在光滑的表面（左）和粗糙的表面（右）反射。

你将需要：一只手电筒。

在你所在的房间里，亮的物体更多还是暗的物体更多？

1 将手电筒作为光源，照向房间里的物体。画出表格，记录你的观察结果。

2 画出你将手电筒照向不同物体的光路图，展示当光到达物体时，它是如何传播的。

How Reflections Work

Light that bounces off an object is called reflected light. If the surface of the object is very smooth and very shiny, most of the light bounces back, or is reflected, in the same direction. All the light enters our eyes and we see a reflection of ourselves.

We can still see objects that are rough, but because the light is scattered, we do not see a reflection. Not all the scattered light enters our eyes.

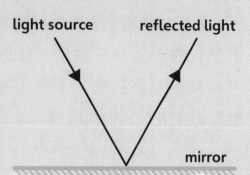

A mirror is a good reflector of light.

What is a mirror?

Look carefully at a mirror. What is it made of? What is on the back? Is the mirror transparent or opaque? Describe to your group how you decided.

Look into a piece of glass, such as a window. Can you see a reflection?

How is the glass different from the mirror?

Mirrors are one example of a type of surface that creates a reflection for us to see ourselves. Glass can make a reflection, but not as well as a mirror. Glass is transparent so some of the light rays travel through the glass and some are reflected.

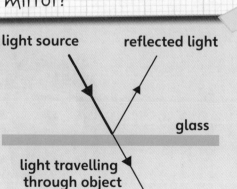

Glass only reflects some of the light, as it is transparent.

反射原理

从物体上反射出的光叫作反射光。如果物体的表面十分光滑、明亮，那么大多数光都会按原路反弹回去或被反射。所有的光进入我们的眼睛，从而让我们看到自己的镜像。

我们也能看到表面粗糙的物体，但是，由于光存在散射，所以我们可能看不到反射光。不是所有的散射光都会进入我们的眼睛。

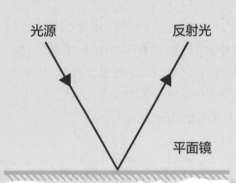

平面镜是很好的反光体。

什么是平面镜？

仔细观察一面平面镜，它是由什么制成的？平面镜的背面是什么？平面镜是透明的还是不透明的？向你的小伙伴描述一下你的看法吧。

看向一块玻璃，例如，一扇窗户。你能观察到镜像吗？

玻璃与平面镜有什么不同？

平面镜是一种典型的能够让我们看见自己镜像的表面。玻璃也能生成镜像，但不如平面镜的效果好。玻璃是透明的，所以部分光线会穿过玻璃而其他光线被反射。

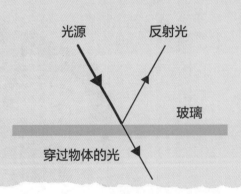

玻璃只会反射部分光线，因为它是透明的。

📄 Glass reflections

You will need: a piece of glass (with safe edges), a piece of black paper, a selection of other materials...

Hold up a piece of glass and look into it. Try to see your reflection. Now hold up a piece of black paper behind the glass and look for your reflection again. What do you notice?

Try different materials behind the glass.

1 Which produces the clearest reflection?

2 Why can you see yourself better in the glass with the black paper behind?

3 Which material placed behind the glass produces the clearest reflection?

When light is reflected, it changes direction. Look at the diagrams on page 21. The light changes direction when it hits the surface of the object.

Mirrors are very useful for reflecting and changing the direction that light travels.

1 Where is the best place to put a mirror if you want to see your reflection?

📄 Hold a mirror in front of you. Look at your reflection. What happens if you move the mirror to one side, or twist it?

Move the mirror so you can see an object behind you. Draw a ray diagram from the source to your eyes to show how you can see the object.

2 How can a mirror help keep you safe if you are driving a car or riding a bicycle?

📖 玻璃反射

你将需要：一块玻璃（边缘不锋利的）、一张黑纸、其他材料……

举起一块玻璃并观察它。尝试着观察你的镜像。然后在玻璃后面放一张黑纸，再尝试着观察你的镜像。你观察到了什么？

尝试在玻璃后面放不同的材料。

1 哪种情况下，你能观察到最清楚的镜像？
2 为什么在玻璃后面放一张黑纸，能让你观察到更清楚的镜像？
3 哪种材料放在玻璃后面能产生最清楚的镜像？

当光发生反射时，它会改变传播方向。观察第21页的示意图。当光到达物体表面时会改变传播方向。

光在传播时，平面镜能很好地反射光并改变光的传播方向。

> **1** 如果你想观察自己的镜像，平面镜的最佳摆放位置在哪里？

📖

将一面平面镜放在你的面前，观察自己的镜像。如果你把平面镜移到一边或者转动平面镜，会发生什么？

将平面镜移动到你能观察到背后物体的位置。画一个光路图，展示光线如何从光源到达你的眼睛并能让你观察到物体的过程。

> **2** 当你在开车或骑自行车时，平面镜如何保证你的安全？

Science Skills

Conclude it! What makes a surface reflective?

Word Box
comparative
conclusion
pattern
variable

📓 Changing surfaces

Take a piece of aluminium foil and hold it up in front of your face. Can you see your reflection in it (it might not be very clear!)?

Fold the foil in half then open it out again carefully. How has your reflection changed?

Fold the foil into four and then open it out flat. What do you notice about your reflection?

Repeat this at least three times, increasing the number of folds.

Draw two ray diagrams to show the difference between the unfolded foil and the foil that has been folded and unfolded several times.

A **conclusion** compares two **variables**. One variable is what we change, the other is what we observe or measure.

📓 **1a** What are the two variables in the Changing surfaces activity?

We write this as a **comparative** sentence to describe the **pattern** of how the two variables are related.

📓 **1b** What is the relationship between these two variables?

2 Explain what the folding of the foil does to the smooth surface.

3 How can you present your observations as evidence to another group? Would photographs be a good idea?

This foil is still shiny, but it no longer shows a clear reflection.

科学技能

得出结论吧！

什么能让物体的表面发生反射？

单词表
comparative 有比较性的
conclusion 结论
pattern 规律
variable 变量

变化的表面

拿出一张铝箔纸，放到你的面前。你能在铝箔纸里观察到自己的镜像吗（镜像可能不太清楚!）？将铝箔纸对折，然后再小心地打开，你的镜像有变化吗？

将铝箔纸对折两次，然后再展开铺平，你观察到你的镜像发生变化了吗？

这些操作至少重复三次，多对折几次。

画出两个光路图，展示一下未被折过的铝箔纸和被对折过多次然后展开的铝箔纸之间的区别。

一个**结论**中包含两个**变量**。一个变量是我们改变的，另一个变量是我们观察或测量的。

1a 在"变化的表面"的实验中，两个变量分别是什么？

我们把结论写成**比较**句，描述了两个变量之间的变化**规律**。

1b 两个变量之间有什么关系？
2 请解释折叠的铝箔纸对其光滑表面的影响。
3 你要如何将你的观察作为证据展示给其他人？拍照片是好的想法吗？

这张铝箔纸还是闪亮的，但已经不再显示清晰的镜像了。

Using Reflection

Mirrors are very useful. Leonardo da Vinci was a famous inventor and scientist living over 500 years ago. He noted many of his ideas down using special writing.

Word Box
periscope

This is how Leonardo da Vinci recorded his notes. What does it say?

Mirror writing

You will need: a marker pen, a piece of paper, a mirror...

Use a thick marker to write the capital letter R on a piece of paper. Place a mirror on the paper next to the letter. What do you notice? Move the mirror to another place on the paper on the other side of the letter. What do you notice? Draw other letters and observe them in the mirror.

Periscopes are very helpful in seeing over the top of things.

Mirrors are also used by sailors in submarines and people in crowds. They help to see over things in an instrument called a **periscope**. A periscope uses two mirrors to reflect light from an object into our eyes.

Periscopes

You will need: two mirrors, string, cardboard or cardboard tubes...

Work with a partner to explore with two mirrors how to see an object on the desk if you are sat underneath the desk. Hold one mirror in front of you and discuss with your partner where the second mirror should be placed so you can see the object. Swap places and try again. Use a piece of string to show the light path from the object to the mirrors and to your eyes.

Apply what you have learned to make a periscope.

运用镜像

平面镜是十分有用的。列奥纳多·达·芬奇是生活在五百多年前的一位著名的发明家和科学家。他将自己的很多想法都用一种特殊的书写方式记录下来。

单词表
periscope 潜望镜

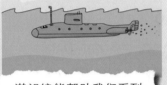

这是达·芬奇记笔记的方式。上面写了什么？

📖 镜像书写

你将需要：一支记号笔、一张纸、一面平面镜……

用一只粗的记号笔在一张纸上写下大写的字母R。将平面镜靠近字母放在纸上。你观察到了什么？再将平面镜移动到字母的另一边。你又观察到了什么？试着写出其他的字母并观察它们在平面镜中的形状。

潜望镜能帮助我们看到潜艇上方的物体。

潜水艇里的水手也会使用平面镜，人们通过平面镜组成的**潜望镜**来观察事物。潜望镜内有两个镜面，用来反射从物体传过来的光，并让光进入我们的眼睛。

潜望镜

你将需要：两面平面镜、线、硬纸板或硬纸管……

和你的小伙伴一起，探索一下，如果你坐在桌子下面，如何用两面平面镜看到桌子上的某个物体。将一面平面镜放在你的前面，再和你的小伙伴一起，讨论一下，另一面平面镜应该放在哪里才能看到物体。让你的小伙伴和你交换一下位置，再尝试一下。试着用一根线表示光从物体传播到镜面，再到你的眼睛所经过的路径。

请运用你学到的知识制作一支潜望镜。

Science Skills

Plan it!

Word Box
control

Reflective clothing helps us to be seen in dim light, or in the dark when a light shines on us, such as from a car headlight. We could wear clothing made of mirror or perhaps made of metal, as metals are also good reflectors.

What properties do clothing materials need to have?

The most reflective material will appear the brightest when a light is shone on it.

Reflective clothing

Work as a group to plan a way to test materials to find the most reflective. Answer these questions to help you with your plan:

1. What variable would you need to change?
2. What variable would you measure?
3. What will you need to **control**, or keep the same in this experiment?

Use the variables you have identified to form the question you are going to investigate.

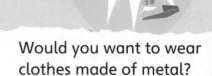

Would you want to wear clothes made of metal?

Reflective clothing helps keep us safe in the dark.

科学技能

做个计划吧!

在光线较暗或漆黑的环境中,穿上带有反光材料的衣服,这样能让我们在被光照射时被其他人看见,例如,来自汽车前灯的光。我们可以穿上由镜子或金属制成的衣服,因为金属也是很好的反光体。

单词表
control 控制

服装的材料需要有哪些特性?

当光照射时,反光性最好的材料是最亮的。

你想穿上由金属制成的衣服吗?

反光的衣服

和你的小伙伴一起做个计划,对不同的材料进行测试并找到反光性最好的材料。回答这些问题可以帮助你制订计划:

1. 你需要改变的变量是什么?
2. 你需要测量的变量是什么?
3. 在实验中,你需要控制或保持不变的变量是什么?

利用你确定的变量来提出你将要探索的问题。

反光的衣服能保证我们在黑暗中的安全。

Safety Clothing

Your clothes can keep you safe. People who work around traffic wear special clothes so that drivers can see them easily. You may have noticed them. The clothes are usually yellow or orange with reflective strips on them.

The reflective strips help the clothing show up when light shines on them when it is dark. Some colours of clothes are more easily seen in daytime or at night.

People working near roads need to be seen easily.

1 Look at these children. Whose clothes are better for cycling at night? Why?
2 Which is the best bike for cycling at night? Why?

A

B

Brightest colour

You will need: pieces of card of different colours, a large piece of white paper or cloth, a torch...

Hold up some different-coloured pieces of card on a white surface on the desk. Shine a torch on the coloured card (this will make a reflection on the white surface). Which colour makes the brightest reflection? Which colour is the least reflective?

3 Why are the clothes that workpeople wear, yellow or orange?
4 Would black be a good colour of clothing to wear at night? Why?

安全衣

你的衣服能保证你的安全。马路上的工人穿着特殊材质的衣服，这样路过的司机能注意到他们。也许你已经注意到他们了，他们的衣服通常是黄色的或橙色的，并且上面还有反光条。

在黑暗中，反光条可以让衣服在有光照过来时被人们注意到。一些颜色的衣服在白天或晚上更容易被看到。

在马路上工作的人需要轻易地让别人看见。

1 观察右图中的儿童，谁的衣服更适合在晚上骑自行车时穿？为什么？

2 哪辆自行车更适合在晚上骑？为什么？

最亮的颜色

你将需要：不同颜色的卡片、一大张白纸或白布、一只手电筒……

把白纸或白布铺到桌面上，手持不同颜色的卡片放在白纸或白布上。对着卡片打开手电筒（光将会在白纸或白布上反射）。哪种颜色的卡片反射最明显？哪种颜色的卡片反射最不明显？

3 为什么工人穿的衣服是黄色的或橙色的？

4 黑色的衣服适合在晚上穿吗？为什么？

Science Skills

Measure it!

We can 'observe' reflections and light, but as scientists we need **evidence** from measurements or from comparisons. Scientists have many pieces of equipment for measuring, including those that measure light.

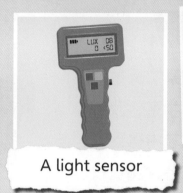
A light sensor

Word Box
evidence
intensity
light sensor
Lux

We can use a **light sensor**, sometimes connected to a computer, to measure the **intensity** of light. The units we measure this in are called **Lux**.

Measuring light

You will need: a selection of fabrics, a torch, a light sensor…

Discuss which fabric you think might be the most reflective. Explain your ideas.

Shine a torch on your fabric. Record the light intensity reflected or make a comparison. Record the data.

1 If you did not have a light sensor, how else could you compare the light reflected from different fabrics?

2 Which would be more accurate, a light sensor or a photograph? Why?

Light intensity can be compared between two materials.

3 Which of these materials is the most reflective? How can you tell?

科学技能

测量一下吧!

我们能"观察"反射和光,但是作为科学家,我们需要从测量或比较中得到**证据**。科学家们有各种测量仪器,包括测量光的仪器。

一个光传感器

单词表
evidence 证据
intensity 强度
light sensor 光传感器
Lux 勒克斯

我们可以使用**光传感器**来测量光的**强度**,有时光传感器是连接到电脑上的。光的强度的单位是**勒克斯**。

📖 测量光

你将需要:不同布料、一只手电筒、一个光传感器……

请讨论一下你认为哪种材料的反光性最好。解释一下你的想法。

用手电筒照向布料。记录反射光的强度并进行比较。记录一下你的数据。

📖 **1** 如果你没有光传感器,还有其他方法可以比较不同布料的反射光的强度吗?

2 哪种方法更准确?光传感器还是照片?为什么?

光的强度能在两种材料之间进行比较。

📖 **3** 哪种材料的反光性最好?你是如何分辨的?

Science Skills

Record it!

Word Box
reliable

Tariq, Bilal and Hiba carried out their experiment on light intensity three times. They had different results each time:

Material	1st reading	2nd reading	3rd reading
Neolyte	32 Lux	31 Lux	32 Lux
Novoflect	25 Lux	23 Lux	25 Lux
Protoflect	10 Lux	11 Lux	12 Lux

1 Should they trust their results? Why?

2 Do you need to repeat your readings for the Measuring light activity on page 33? Why?

Repeating an experiment allows us to check our results. If the results are all similar we can trust the results. Scientists say that the results are **reliable**.

There are different ways we can record results. We could draw a table and record data, or place photos in a table for comparison. We could draw a graph of our data. Bar charts are used when we have names for objects as one of the variables. Line graphs are used when we have numbers for both variables.

3 Look at this table of data. Why is this graph the correct one to draw?

Material	Light reading
Neolyte	32 Lux
Novoflect	25 Lux
Protoflect	10 Lux

Graph to show light intensity of materials (Lux)

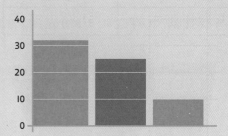

4 What is missing from this graph?

科学技能

记录一下吧！

塔里克、比拉尔和希巴进行了三次关于光的强度的实验。每次结果都不一样：

材料	第一次的数据	第二次的数据	第三次的数据
新石牌布料	32勒克斯	31勒克斯	32勒克斯
诺沃牌布料	25勒克斯	23勒克斯	25勒克斯
普罗托牌布料	10勒克斯	11勒克斯	12勒克斯

单词表
reliable 可靠的

1 他们应该相信这些结果吗？为什么？

2 你需要再次读取第33页测量光的实验中的数据吗？为什么？

重复一个实验能让我们检查实验结果。如果每次结果相似，那么我们就能信任实验结果。科学家们就称实验结果是**可靠的**。

我们有不同的方式来记录结果。可以通过画表格来记录数据，或将图片放到表格中做对比。

我们可以把数据画成各种统计图。条形图可以用不同名称代表变量。折线图可以用数字来代表变量。

材料	光的强度
新石牌布料	32勒克斯
诺沃牌布料	25勒克斯
普罗托牌布料	10勒克斯

3 观察表格中的数据。为什么这张条形图是正确的？

不同材料的光强的条形图（勒克斯）

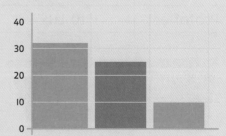

4 这张条形图上缺少什么？

Making Shadows

To make a shadow we need to have a light source and an opaque object.

1 Why does the object need to be opaque?

Shadow shapes

Hold your hand in front of a light source. Can you change the shape of your hand to produce a shadow shape that looks like something else? Can your friends recognise this new shadow shape?

What shape has this pair of hands made?

 2a How do you think a shadow is made?

 2b Draw a ray diagram to help explain your ideas.

People have used shadows to produce pictures for many hundreds of years. There is a special name for pictures made from shadows. They are called silhouettes. We can describe a silhouette as the outline of something, without any details of the object.

Silhouettes were popular before photography was invented.

3 Look at these diagrams. Are they silhouettes? Why?

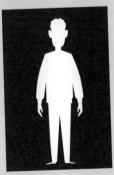

制作影子

为了制作影子，我们需要一个光源和一个不透明的物体。

1 为什么需要不透明的物体？

影子的形状

把你的手放到光源前，你能通过改变双手的形状来制作一个看上去像别的事物的影子吗？你的朋友们能认出新影子的形状吗？

这双手摆出的影子像什么？

2a 你认为影子是如何形成的？

2b 画一张光路图来帮助你解释你的想法。

人们利用影子来制作画，这已经有几百年的历史了。这种画作有个特殊的名字，叫作轮廓画。轮廓是指物体外部的形状，而不包含任何细节。

在摄影技术未被发明以前，轮廓画十分受欢迎。

3 观察右边的图片。它们是轮廓画吗？为什么？

A group of younger children have been exploring shadows. They were outside in the playground, using the Sun as their light source. They have drawn some pictures of the shadows they made.

4 Explain why each shadow is not correctly drawn.

5 Some of the children noticed that their shadows changed during the day. The shadows moved across the ground and got longer then shorter. Why do you think this is?

一群小朋友在探索影子。他们在室外操场上，太阳是他们的光源。他们画出了自己的影子。

📔 4 请解释一下为什么每个影子都画得不正确。

📔 5 有些小朋友注意到他们的影子会在一天中发生变化。投射在地上的影子先变长后变短。你认为这是为什么呢？

Science Skills

Predict it!

A shadow is made when an object blocks light from travelling. We can change the shadow by changing the variables in how the shadow is made.

Read the ideas and predictions that the other children have said about how the shadow can change.

If you move the lamp nearer to the screen the shadow will get smaller.

If you move the lamp the shadow will move.

The shadow is the biggest when the puppet is closest to the lamp.

I do not think it matters where the lamp is.

The shadow will get bigger if you move the puppet closer to the screen.

For a good prediction you should try to explain the science behind your ideas.

1 Which predictions do you think are correct? Why?

2 Which of the pieces of information given here, helps to explain the children's ideas above?

- Light travels in straight lines.
- Light travels very fast, faster than sound.
- Opaque objects block light to make shadows.
- Light travels from a source, bounces off an object and into our eyes.

科学技能

预测一下吧!

当物体阻挡光传播时,就会产生影子。通过改变影响影子产生的变量,我们可以改变影子。

阅读其他小朋友们的关于改变影子的观点和预测。

- 如果你把灯放在离屏幕更近的地方,影子会变小。
- 如果你移动灯,影子也会随之移动。
- 皮影娃娃离灯最近的时候,影子最大。
- 我认为灯的位置会影响影子。
- 如果你把皮影娃娃放得离屏幕更近,影子就会更大。

为了得到一个可靠的预测,你应该尝试着解释一下你的观点背后的科学原理。

1 你认为哪些预测是正确的?为什么?

2 右边给出的信息中,哪几条可以用来解释上述小朋友们的观点?

- 光沿直线传播。
- 光传播得很快,比声音传播得更快。
- 不透明物体阻挡光线传播,从而形成影子。
- 光从光源发出,在物体表面反射,最终进入我们的眼睛。

Science Skills

Plan it!

We can explore how to change shadows using a simple equipment. Although this is not a 'fair test' experiment, we will still have variables to identify.

Exploring shadows

You will need: a pencil, modelling clay (or something similar to hold the pencil), a compass template, a light source...

Using the equipment, make a shadow. Describe to your partner how you have made your shadow.

Try these activities and make notes about what you do to change the shadow each time:

1 Make the shadow longer.
2 Make the shadow shorter.
3 Make the shadow point north.
4 Make a short shadow that points south.
5 Make a very long shadow that points west.

1 Answer these questions about the Exploring shadows activity.
  a What variables are there in your exploration?
  b Which variable did you change?
  c Which variable did you observe or measure?
  d Which variables did you control (keep the same)?

2 Look back at the predictions made by the other children. Which predictions are correct?

Shadow instructions

Produce a set of instructions for a younger child so they can make a shadow for themselves.

科学技能

做个计划吧!

我们可以用一个简单的设备来探究如何改变影子。尽管这个探究不是正规的实验,我们仍然需要确定变量。

探究影子

你将需要:一支铅笔、黏土(或类似的可以支撑铅笔的东西)、一个指南针样板、一个光源……

使用上述器材来制作影子。向你的小伙伴描述一下你是如何制作影子的。

尝试下列操作并做好笔记,记录每次你是如何改变影子的:

1 让影子变长。
2 让影子变短。
3 让影子指向北方。
4 制作一个较短的影子并指向南方。
5 制作一个特别长的影子并指向西方。

📓 1 回答探究影子活动里的相关问题。
  a 你的探究中有哪些变量?
  b 你改变的是哪个变量?
  c 你观察或测量的是哪个变量?
  d 你控制(或保持不变)的是哪个变量?

📓 2 回顾其他小朋友做出的预测。哪些预测是正确的?

影子的指示说明

请列出一组指示说明,来帮助更小的小朋友自己制作影子。

Science Skills

Graph it!

Graphing shadows

Remember that we, as scientists, need evidence from experiments. A younger group of children have gathered evidence about shadows. They have produced some data in a table.

It is easier to identify a pattern if the data is turned into a graph. Draw a graph of the data shown in the table below.

Distance from light (cm)	Height of shadow (cm)
10	39
20	33
30	31
40	30
50	29
60	28

1. What is the relationship between the distance from the light source and the height of the shadow?
2. How else can you make a shadow taller?

科学技能

画个图表吧!

画出影子图表

请牢记,作为小科学家,我们需要从实验中得出证据。一组年龄更小的小朋友们已经收集了一些关于影子的证据,他们已经在表格中生成了数据。

如果数据被绘制成了图表,我们就能更容易地找出规律。运用下面表格中的数据画一个图表吧。

与光源的距离 (厘米)	影子的高度 (厘米)
10	39
20	33
30	31
40	30
50	29
60	28

1 物体与光源的距离和影子的高度之间有什么关系?
2 你还有其他方法能增加影子的高度吗?

How Shadows Are Made

To make a shadow we need to block light. We can **model** this with water. When we use a model, we use something familiar that we can see to act like something we cannot see. Models help us to understand things that are difficult to observe.

Word Box
model

Modelling light - the water model

You will need: coloured sugar paper, water in a spray bottle...

Work with a partner. Hold a piece of coloured sugar paper up. Place your hand against the paper. Ask your partner to spray water onto your hand. Remove your hand from the paper.

What do you see when you take your hand away from the paper? Why?

Try moving your hand further away from the paper and spraying with water. What happens? Is this what you expected?

 Explain how the water in the activity above, models light. Is the water sprayer a good model for light?

We can bring all our information about light together to help explain how shadows are made.

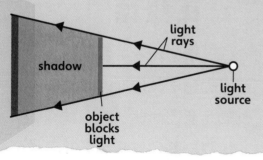

Shadows are formed when an opaque object blocks light from a source.

Comparing ideas

Look back at your ideas about how a shadow is formed. Compare them to the diagram here. Discuss any differences with your group.

影子是如何形成的

为了制作影子，我们需要阻挡光。我们可以用水来制作一个**模型**。当我们运用一个模型时，我们利用能看见的且熟悉的事物来演示我们看不见的事物。模型能帮助我们认识难以用肉眼观察到的事物。

单词表
model 模型

用模型展示光——水模型

你将需要：带色的糖纸、装水的喷雾瓶……

和你的小伙伴一起，拿起一张带颜色的糖纸。将手贴到糖纸上。让你的小伙伴将水喷向你的手。你再把手从糖纸上拿开。

把手从糖纸上拿开后，你观察到了什么？为什么？

试着把手移动到离糖纸更远一点儿的距离，再喷水试一试。你观察到了什么？这在你的预料之中吗？

📖 请解释一下上述实验中水是如何模拟光的。喷雾瓶中的水是恰当的光线传播模型吗？

我们可以收集所有关于光的信息，进一步帮助我们解释影子是如何形成的。

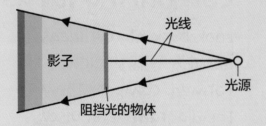

当不透明物体阻挡光源发出的光传播时，影子就形成了。

对比想法

请回顾一下影子是如何形成的想法，把你的想法同左边的示意图对比一下。和你的小伙伴一起，讨论一下，二者之间有什么不同。

Light and Sight

Part 2 科学虫大闯关

1. What two things do we need to be able to see an object?

   _____
   _____

2. Discuss with your group how do you think these two things work together to help us see.
   Draw your ideas in the space below.

All about Light Sources

1. How many different light sources can you think of? Use the Venn diagram below to sort your light sources into human-made (artificial) or natural sources of light.
   Are there any light sources which could be both, or could be neither? Where would you place these on the diagram?

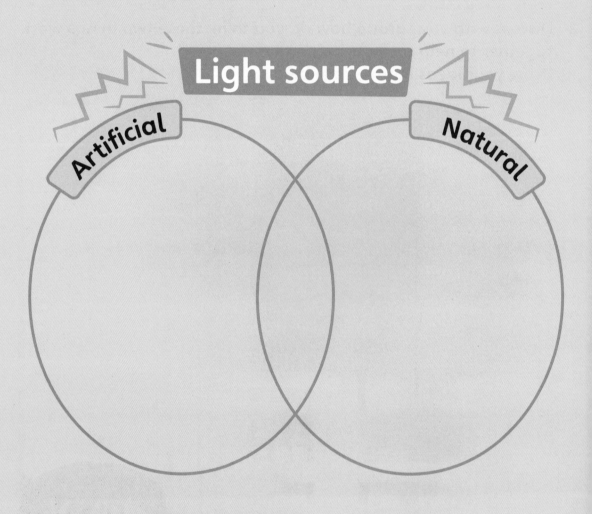

2. Which list is longer? Why is this?

3. Look at all the light sources you have named and the items in the diagram below. Decide if each of these objects is a source of light or reflects light, then decide if it is human-made (artificial) or natural. Write the name of the object into the Carroll diagram below. Include any others you can think of. The first one has been done for you.

	Human-made	Natural
Light source	Table lamp	
Reflector		

4 Do you notice anything interesting about the groups in the Carroll diagram on page 51? For example, which group is the largest? Which is the smallest? What ideas do you have about why this is?

5 Look at all the light sources and compare them to the reflectors. What definition could you give to a light source, so it is not confused with an object that only reflects light?
Hint: What does the light source do, that the reflector does not do?

6 Use this definition to help you explain why a mirror is not a light source.

Travelling Light

You will need a calculator to work out these answers, and the following information:

Light speed = 300,000 kilometres per second
Distance of Sun from the Earth = 148 million kilometres
1 year = 365 days, 1 hour = 60 minutes, 1 minute = 60 seconds
Time for light to travel around the Earth seven times = 1 second

Show your workings out in the spaces below the questions, as well as the answer.

1   Calculate how long it takes light from the Sun to reach Earth.

2   Calculate the circumference of the Earth, using light speed and how long it takes to travel around the Earth seven times.

3   A light year is the distance light can travel in a single year. Calculate how far this is, in kilometres.

Light Rays

Use these activities to show the idea that light rays travel in straight lines. Explain what you can see and how they prove the theory.
Carry out part 1 of the activity on page 9 in Part one.

1. Why can you not see anything when the cardboard tube is bent?

_____
_____
_____

Carry out the part 2 of the Ray of light activity in Part one.

2. What can you see on the black card? Why?

_____
_____

3. What happens when you lift up one of the pieces of card in the line? Why?

_____
_____

Shine a torch through a comb.

4. What can you see? Why?

_____
_____

Light News

1. Write a short newspaper style report as if you had just discovered how light travels.
   Here are some hints to help write a news report:

   > CAPTURE INTEREST WITH AN EXCITING HEADLINE
   >
   > Sub heading – Sum up what the report is about in the first sentence
   >
   > *By line – put your name to the article*
   >
   > Lead paragraph – Describe what the article is about. Write in the past tense.
   >
   > Body of report – Continue the story, include facts. Write in the past tense. Include facts about what happened, how it happened and why it happened.
   >
   > Use a labelled diagram to help tell more of the story.
   >
   > Conclusion – Summary of your report at the end.

Light for Seeing

Roll up a piece of black paper into a tube. Place the tube over a small object, such as a coin or a paperclip. Look down the tube.

1 **What can you see? Why?**

_____

Explore making holes in the side of the tube. (Be careful if you are using scissors to make the holes.) Look down the tube each time you make or change a hole.

2 **What is different about what you see when looking down the tube now?**

_____

3 **How does where you make the hole affect how well you can see? Why do you think this is?**

_____

Try shining a torch through one of the holes. Point the torch downwards towards the object, then upwards towards your eyes.

4 **Which direction allows you to see the object most easily?**

_____

Light Paths

Look at the ray diagrams showing light paths. What is wrong with each of them?

Next to each ray diagram, write how you would correct it.

1

_____
_____

2

_____
_____

3

_____
_____

The Eye

1. Use a range of sources to find out about Alhazen, also known as Al-Hasan Ibn al-Haytham, an Arabian scientist who was the first to explain how we see.
   Use the information you gather to produce a fact file on him.
   Share this with your group.

2 Read the statements below. They describe the way we see things but are not in the correct order. Decide what order they should be in. Write the numbers 1–10 to place the statements in the correct order.

- ____ Light radiates from the source.
- ____ Our brain interprets the light as an image of the object.
- ____ A message is transmitted along the optic nerve to the brain.
- ____ The reflected light travels very fast in a straight line towards our eyes.
- ____ Light hits the object.
- ____ Light hits the light sensitive cells at the back of the eye.
- ____ The light reflects from the surface of the object and is scattered.
- ____ Light travels very fast in straight lines.
- ____ The brain interprets the message so we see the object.
- ____ Light enters our eye through the black hole in the middle, called the 'pupil'.

3 What do you think the purpose of your pupil is? Hint: How does it protect your eyes? Name another part of your eye that protects it.

How We See

1. Are there more shiny or dull objects in the room where you are? Use a torch as your light source to shine light on objects in the room.
   Draw a table in the space below to record your observations.
   Hint: How many columns will you need?

2. Draw at least two ray diagrams of the objects you shone the torch on, using light rays (with arrows) to show how the light behaves when the light hits the object.

How Reflections Work

Carry out the Glass reflections activity on page 23 in Part one.

1a Why can you see yourself better in the glass with the black paper behind?

_____
_____

1b Complete this table to show which materials produced the clearest reflection when placed behind the glass. For each material, tick whether it produced a good or a poor reflection.

Material	Good reflection	Poor reflection

2 Which material placed behind the glass produced the clearest reflection? Why do you think this is?

_____
_____

3 Draw the light rays with arrows on to complete the diagram below to show how you can see an object behind you, using a mirror.

Conclude it!

Carry out the Changing surfaces activity on the page 25 in Part one.

1. Describe your observations of your reflection in the unfolded foil and in the folded then unfolded foil.

2. What are the two variables in this activity?

   Variable I changed: _____

   Variable I observed: _____

3. How are these two variables related? Write a comparative sentence.

4. Draw a ray diagram or write an explanation, to finish your conclusion.

Using Reflection

Place a mirror next to each of these letters, so you can see its reflection.

R   S   T   A   b   p

1. Draw each of the letters you see in the mirror in the space below each letter. What do you notice?

2. Use the space below to write your name in mirror writing.

3. Explain why the words 'Police' and 'Ambulance' might be written back to front on the front of their vehicles.

4. Look at this writing. Which of these words is written with a mirror?

A noitɔelʇeɿ

B noitɔolʇeɿ

C ɿeʇlectɔion

D noitɔelʇeɿ

Science Skills

Plan it!

You are investigating which material is the 'best' reflector to make clothing out of.

1. Describe what the word 'best' means in this experiment?
   Hint: is it the most reflective or the least?

2. List all the variables that could change or be measured in this experiment:

Variables I could change	Variables I could measure or observe

3. Highlight or circle the variables you will experiment with.

4. What will you do with the other variables?

5. Use the following format to produce a testable question for your experiment:
   How does *what I change* affect *what I would observe or measure*?

6. What will you use to measure or observe?

Science Skills

Measure it!

Carry out the Measuring light activity on the page 33 in Part one.

1. If you did not have a light sensor, how else could you compare the light reflected from different fabrics?

2. Which would be more accurate of the methods you have available? Why?

3. How will you record your results? Circle at least one of the diagrams.

4. Draw your own method for recording and fill in your data in the space below.

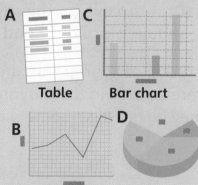

Table   Bar chart
Line graph   Pie chart

5. Write a sentence to describe what you found out about which is the most reflective material.

Making Shadows

1. Use the space below to draw a ray diagram of how you think a shadow is made. The screen, object and light source have been drawn for you. Add in the light rays and arrows.

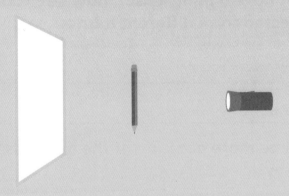

2. Describe in words how you think the shadow is made. Use the following key words: light source, light ray, blocks, travels, opaque.

3. Draw an eye on the ray diagram above to show where we should place our head to be able to see the shadow.

A group of younger children have been exploring shadows. They were outside in the playground, using the Sun as their light source. They have drawn some pictures of the shadows they made.

A

4 Look at the children's shadow drawings. Explain why each of the children's shadows is not drawn correctly.

5 Some of the children noticed that their shadows changed during the day. The shadows moved across the ground and got longer then shorter. Why do you think this is?

Science Skills

Plan it!

Explore making shadows on the compass template. Carry out the Exploring shadows activity on the page 43 in Part one.

1. List the variables in your exploration. A variable is something that you can change, or something you observe or measure.

   _____
   _____
   _____
   _____

2. Circle the variable you changed in blue. Circle the variable you observed or measured in green. Underline the variables you kept the same.

3. Which two of these ideas are correct? Circle them and explain why you think this is correct.
   'If you move the lamp the shadow will move.'
   'If you move the lamp nearer to the screen the shadow will get smaller.'
   'The shadow will get bigger if you move the puppet closer to the screen.'
   'The shadow is the biggest when the puppet is closest to the lamp.'
   'I don't think it matters where the lamp is.'

   _____
   _____
   _____

Science Skills

Graph it!

A younger group of children have been exploring shadows as well. They have produced some data in a table.

1 Help the children to see a relationship, or pattern, in their data by drawing a graph for them.

Distance from light (cm)	Height of shadow (cm)
10	39
20	33
30	31
40	30
50	29
60	28

Remember: a graph needs a title, labels and units on each axis. Each axis should also be evenly spaced and use as much of the graph paper as possible.

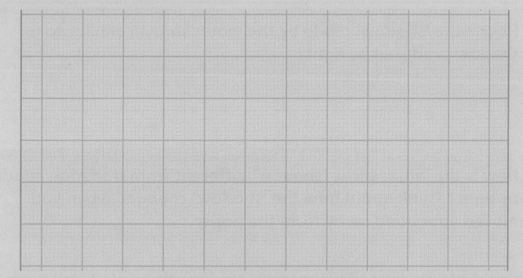

2 What is the relationship, or pattern, between the distance from the light source and the size of the shadow? (Write this as a comparative sentence.)

_____
_____

How Shadows Are Made

Use the questions below to share your ideas about the water model.

1. How is the water sprayer similar to a light source?

2. What does the water sprayer represent?

3. Can the water spray travel through your hand? Why?

4. How is the 'shadow' made by the water sprayer similar to a shadow produced by light?

5. How is the water spray 'shadow' different from a shadow made by light? Think about how the 'shadow' changed when you moved your hand away from the paper.

6. Do you think the water sprayer is a good model or not? Why?

True or False?

There are many facts about light that you have learned. Which of these statements made by other children are true and which are not true? Tick true or false.

	True	False
Light travels in straight lines.		
Light travels faster than sound.		
Light enters our eyes so we can see.		
The role of our pupils is to control the amount of light entering our eye.		
The pupil is the colour part of the eye.		
A light source is bright and shiny.		
Light can be bent.		
When light bounces from a smooth shiny surface we see a reflection.		
Shadows have the same shape as the object blocking the light.		
Opaque objects produce the darkest shadows.		
Moving an object closer to the light source produces a bigger shadow.		
There are more artificial light sources than natural light sources.		
Rough surfaces scatter light.		
Black materials are good reflectors of light.		
Light reflects from all objects.		

Light Story

Use the space on this page to plan a six-page booklet of what you have learned about light. Ideas might include:
- How light travels
- How we see
- What a reflection is
- How to use mirrors
- How to make a shadow
- How to change a shadow

Glossary

artificial: not natural; human-made

comparative: comparing two variables together

conclusion: explain what has been discovered with reasons

control: keep the same as part of an experiment

evidence: the proof of something

intensity: how strong or bright something is

interpret: to turn information into another form that is more usable

light-sensitive: response to the level (amount or intensity) of light

light sensor: a device for measuring light intensity

light speed: how fast light travels

Lux: the unit of measurement of light intensity

model: a representation of something that helps explain it more simply

optic nerve: transmits messages from the eye to the brain

pattern: a relationship which repeats between two or more variables

periscope: an object using mirrors to look over the top of something

pupil: the black part in the centre of the eye

radiate: spread out in all directions

ray diagram: scientific diagrams to show light rays travelling

ray: beams of light that travel in straight lines

reflection: when light bounces off a shiny flat surface; what we see when we look into a mirror

reflect: light 'bouncing' off a surface; sound can also reflect (called an echo)

reliable: something that can be trusted

scatter: move quickly in different directions

transmit: move from one place to another

variable: things that can change

词汇表

人造的：非自然的；人工制造的

有比较性的：同时比较两个变量

结论：用原因解释发现的结果

控制：保持实验的某个部分不变

证据：证明事件成立的事实

强度：描述物体的强烈程度或亮度

解释：将信息以更实用的形式传达

感光的：能对不同强度的光做出回应

光传感器：测量光的强度的仪器

光速：光传播的快慢

勒克斯：光强的单位

模型：仿照某事物制成的，帮助解释某事物的原理

视觉神经：可以把眼睛获得的信息传递给大脑

规律：两个或两个以上的变量之间的关系

潜望镜：借助镜子看到上方的物体

瞳孔：眼睛中间黑色的部分

辐射：在不同的方向传播

光路图：展示光线传播路径的科学示意图

光线：沿直线传播的光束

镜像：光在光滑表面上的反弹；我们在镜子中看到的事物

反射：光在物体表面上发生"反弹"的现象；声音也可以反弹（称为回声）

可靠的：可以被信任的事物

散射：在不同方向快速地分散开

传递：从一个地方传到另一个地方

变量：可以改变的事物

培生科学虫双语百科

奇妙物理

Changing Circuits

变化的电路

英国培生教育出版集团 著·绘

徐昂 译

电子工业出版社
Publishing House of Electronics Industry
北京·BEIJING

Original edition, entitled SCIENCE BUG and the title Changing Circuits Topic Book, by Debbie Eccles published by Pearson Education Limited © Pearson Education Limited 2018
ISBN: 9780435195434

All rights reserved. No part of this book may be reproduced or transmitted in any form or by any means, electronic or mechanical, including photocopying, recording or by any information storage retrieval system, without permission from Pearson Education Limited.

This adaptation of SCIENCE BUG is published by arrangement with Pearson Education Limited. Chinese Simplified Characters and English language (Bi-lingual form) edition published by PUBLISHING HOUSE OF ELECTRONICS INDUSTRY, Copyright © 2023.

For sale and distribution in the mainland of China exclusively (except Hong Kong SAR, Macau SAR and Taiwan).

本书中英双语版由 Pearson Education（培生教育出版集团）授权电子工业出版社在中华人民共和国境内（不包括香港、澳门特别行政区及台湾地区）独家出版发行。未经出版者书面许可，不得以任何方式抄袭、复制或节录本书中的任何部分。

本套书封底贴有 Pearson Education（培生教育出版集团）激光防伪标签，无标签者不得销售。

版权贸易合同登记号　图字：01-2022-2381

图书在版编目（CIP）数据

培生科学虫双语百科. 奇妙物理. 变化的电路：英汉对照 / 英国培生教育出版集团著、绘；徐昂译. --北京：电子工业出版社，2024.1
ISBN 978-7-121-45132-4

Ⅰ.①培… Ⅱ.①英… ②徐… Ⅲ.①科学知识-少儿读物-英、汉 ②电学-少儿读物-英、汉 Ⅳ.①Z228.1 ②O441.1-49

中国国家版本馆CIP数据核字（2023）第035084号

责任编辑：李黎明　文字编辑：王佳宇
印　　刷：河北迅捷佳彩印刷有限公司
装　　订：河北迅捷佳彩印刷有限公司
出版发行：电子工业出版社
　　　　　北京市海淀区万寿路173信箱　邮编：100036
开　　本：787×1092　1/16　印张：35　字数：840千字
版　　次：2024年1月第1版
印　　次：2024年2月第2次印刷
定　　价：199.00元（全9册）

凡所购买电子工业出版社图书有缺损问题，请向购买书店调换。若书店售缺，请与本社发行部联系，联系及邮购电话：（010）88254888，88258888。
质量投诉请发邮件至zlts@phei.com.cn，盗版侵权举报请发邮件至dbqq@phei.com.cn。
本书咨询联系方式：010-88254417，lilm@phei.com.cn。

使用说明

欢迎来到少年智双语馆！《培生科学虫双语百科》是一套知识全面、妙趣横生的儿童科普丛书，由英国培生教育出版集团组织英国中小学科学教师和教研专家团队编写，根据英国国家课程标准精心设计，可准确对标国内义务教育科学课程标准（2022年版）。丛书涉及物理、化学、生物、地理等学科，主要面向小学1~6年级，能够点燃孩子对科学知识和大千世界的好奇心，激发孩子丰富的想象力。

本书主要内容是小学阶段孩子需要掌握的物理知识，含9个分册，每个分册围绕一个主题进行讲解和练习。每个分册分为三章。第一章是"科学虫趣味课堂"，这一章将为孩子介绍科学知识，培养科学技能，不仅包含单词表、问题和反思模块，还收录了多种有趣、易操作的科学实验和动手活动，有利于培养孩子的科学思维。第二章是"科学虫大闯关"，这一章是根据第一章的知识点设置的学习任务和拓展练习，能够帮助孩子及时巩固知识点，准确评估自己对知识的掌握程度。第三章是"科学词汇加油站"，这一章将全书涉及的重点科学词汇进行了梳理和总结，方便孩子理解和记忆科学词汇。

2024年，《培生科学虫双语百科》系列双语版由我社首次引进出版。为了帮助青少年读者进行高效的独立阅读，并方便家长进行阅读指导或亲子共读，我们为本书设置了以下内容。

（1）每个分册第一部分的英语原文（奇数页）后均配有对应的译文（偶数页），跨页部分除外。读者既可以进行汉英对照阅读，也可以进行单语种独立阅读。问题前面的 📖 符号表示该问题可在第二部分预留的位置作答。

（2）每个分册第二部分的电子版译文可在目录页扫码获取。

（3）本书还配有英音朗读音频和科学活动双语视频，也可在目录页扫码获取。

最后，祝愿每位读者都能够享受双语阅读，在汲取科学知识的同时，看见更大的世界，成为更好的自己！

电子工业出版社青少年教育分社
2024年1月

Contents 目录

Part 1　科学虫趣味课堂　　　　　　　　　　　/ 1

Part 2　科学虫大闯关　　　　　　　　　　　　/ 47

Part 3　科学词汇加油站　　　　　　　　　　　/ 71

Part2译文

配套音视频

Changing Circuits

Many things that we use every day need electricity to work. For example, a television and mobile phone both need electricity. The television usually uses mains electricity, and the mobile phone uses a cell.

The terms battery and cell sometimes get mixed up. A cell is a single electrical energy source that converts chemical energy into electrical energy. A battery is two or more cells working together. Mains electricity is the supply from power stations to our homes.

Unfortunately, electrical appliances are not always reliable. There are many reasons why they can stop working. This torch is not working.

1. Identify the cell, bulb and switch of this broken torch.
2. What could you do to try to fix it?
3. List the reasons why the torch may not be working.
4. Draw the electrical components needed to make a working torch.

变化的电路

日常生活中我们使用的很多东西都需要用电才能工作。例如,电视机和手机都要用电。电视机利用的是干线供应的电力,手机利用的是电池。

术语电池(cell)和蓄电池(battery)有时会被人们弄混。电池指的是单个的能够将化学能转化为电能的能源装置,而蓄电池则是由两节或多节电池串联在一起工作的。干线供应的电力是从发电站直接运往家中的电能。

然而,电器并非永远都可靠,有很多原因能让电器停止工作。这个手电筒就不能正常工作。

1. 指出这个损坏的手电筒的电池、灯泡和开关。
2. 你要如何做才能修好手电筒?
3. 列出导致手电筒不能正常工作的原因。
4. 画出手电筒正常工作需要的电路元件。

Circuit Diagram Symbols

Symbols are simple marks or signs that represent something. For example, in mathematics the + symbol is used to represent add. All the letters in our alphabet are also symbols. As they are simple, symbols are quick to draw and easy to understand.

Word Box
circuit diagram
symbol

Engineers, electricians and scientists all use the same **circuit diagram symbols** for different components of electrical circuits. They do this so they can give information about the circuits to others, and to avoid making mistakes such as using the wrong component when they are constructing a circuit.

This circuit diagram is the type electricians use in their work. They use them to understand how an electrical appliance works and to find faults.

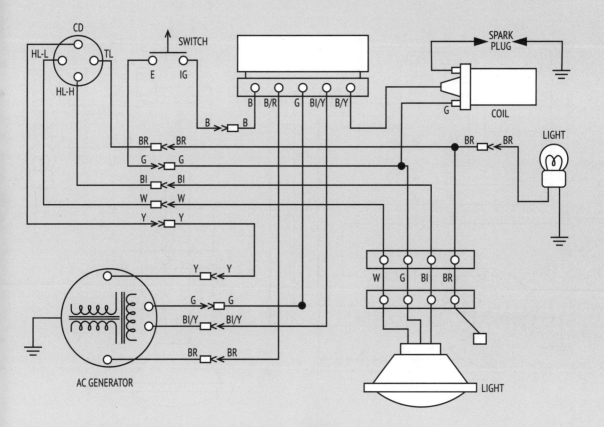

电路图符号

符号是代表事物的简单标识。例如，数学中的"+"表示相加，英文中的字母表同样也是符号。因为符号比较简单，所以人们可以快速地画出来并容易理解。

工程师、电气技师和科学家们在不同的电路图中运用相同的**电路图符号**。他们可以通过这种方式将电路的信息传达给别人，避免出错，例如，在搭建电路时可以避免使用错误的电路元件。

这个图是电气技师在工作时会用到的典型电路图。他们运用电路图来理解某种电器运行的原理并找出有问题的地方。

> **单词表**
>
> circuit diagram symbol
> 电路图符号

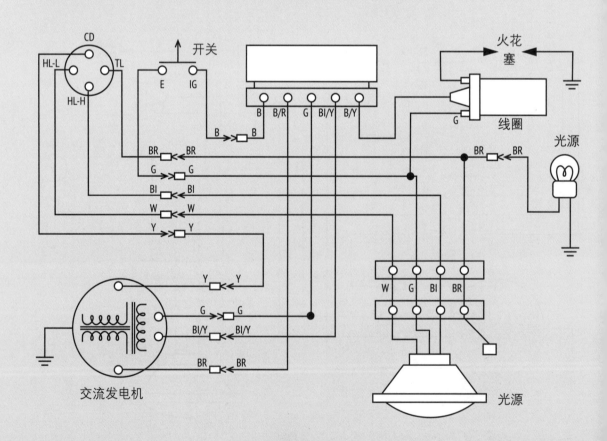

Here are some symbols used in circuit diagrams.

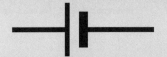 

Match components to their circuit diagram symbols

You will need: a cell, buzzer, bulb, motor, switch and wire...

Work in a pair.

1. Match the component to its circuit diagram symbol.
2. Decide whether you think each symbol works well.
3. Discuss any changes you would make to improve any of the symbols. List these and explain your reasons.
4. Choose one of the symbols that you think could be improved and draw it. Then draw your improved symbol. Explain why you think this is an improvement.

Circuit diagram symbols are the same all over the world. This means we cannot change the symbols.
Why is this sensible?

这些是电路图中会用到的一些符号。

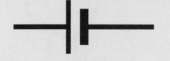

匹配电路元件和电路图符号

你将需要： 电池、蜂鸣器、灯泡、电动机、开关和导线……和你的小伙伴一起行动。

1. 匹配电路元件与电路图符号。
2. 判断一下每个电路图符号是否准确地代表了电路元件。
3. 讨论一下，你们是否想对某些符号做出一些改动，请列出来并解释一下。
4. 选择一个你们想优化的某个电路图符号并画出来。然后画出你们改动后的电路图符号，并解释为什么认为这样的改动是一种优化。

> 电路图符号在全世界通用，这意味着我们不能更改任何符号。为什么现在的电路图符号是合理的呢？

Remembering Symbols

Now that you have studied the circuit diagram symbols you should be able to remember what each one looks like.

 1 Close your book. Draw the symbols.
2 Check them using the pictures you have studied.

Describing circuit diagram symbols

You will need: 12 identical 5x5 cm blank cards, a pencil, a two-minute timer...

Work in a pair.

1 Take six cards each.
2 Each draw a different circuit diagram symbol on each piece of card. This means that you will have two of each type of circuit diagram symbols in total. (Two cells, two bulbs, two buzzers, two motors, two switches and two wires.)
3 Shuffle all the cards together so you do not know what order they are in. Then place the cards face down in a pile between you.
4 The rules are that one of you will take the top card, but do not show it to your partner or name the symbol. Describe the symbol, and your partner guesses which it is.
5 Set the timer for two minutes. Take it in turns to pick a card from the top of the pile. Describe the symbol. As soon as your partner correctly guesses it place it face up. See how many you can identify in two minutes.

记忆符号

既然你已经学习了一些电路图符号，现在你应该能记住每个符号了。

> 1 合上书，将这些符号画出来。
> 2 结合你刚才学习的图片，检查一下。

描述电路图符号

你将需要：12个相同的5厘米×5厘米的空白卡片、一支铅笔、能计时两分钟的计时器……

和你的小伙伴一起行动。

1 每个人拿六张卡片。

2 两个人分别在每张卡片上画一个不同的电路图符号，这意味着你们一共拥有两套电路图符号。（两节电池、两个灯泡、两个蜂鸣器、两个电动机、两个开关和两根导线。）

3 将卡片随机打乱，这样你们就不知道卡片排列的顺序了。然后将所有卡片正面朝下，整齐地堆放在你们之间。

4 规则：你们其中的一个人拿走最上面的卡片，但是不要给另一个人看，也不要告诉另一个人卡片上的符号名称。你对符号进行描述，让你的小伙伴来猜是什么。

5 计时两分钟。轮流来拿最上面的卡片，再对其进行描述。一旦你的小伙伴猜出正确的符号名称，就将卡片正面朝上摆放，看看你们在两分钟内能猜出多少个电路图符号。

Science Skills

Complete Circuits – Record it!

Word Box
complete circuit
series circuit

We need a complete loop for the components in an electrical circuit to work. We call this a **complete circuit**. If there is a break in the circuit it will not work. When the components are part of the same complete loop this is known as a **series circuit**.

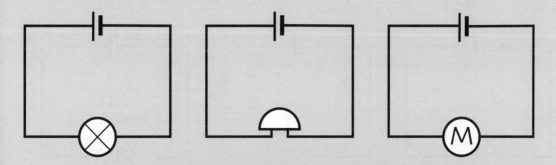

Look at these circuit diagrams and predict what will happen in each of them.

Using circuit diagrams to construct circuits

You will need: wires, a cell, a buzzer, a bulb, a motor...

Work in a small group.

1. Make the circuits shown in the diagrams.
2. Did they do what you expected?
3. If you could put a switch in any of the circuits which would you put the switch in? Why?
4. What components are common to all the circuits?

科 学 技 能

闭合电路——记录一下吧!

单词表

complete circuit 闭合电路
series circuit 串联电路

电路原件需要构成闭合回路才能保证电路正常运行,我们把这样的电路叫作**闭合电路**。如果电路中的某处断开,则电路就不能正常运行。如果电路元件在同一个闭合回路中,则该电路就可以称为**串联电路**。

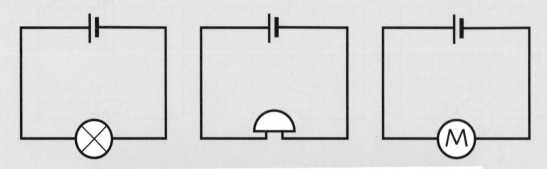

观察上面的电路图,预测一下每张电路图会发生什么。

借助电路图来搭建电路

你将需要:若干导线、一节电池、一个蜂鸣器、一个灯泡、一个电动机……

和你的小伙伴一起行动。

1 根据电路图搭建电路。

2 电路的效果符合你的预期吗?

3 如果你在任意的电路中添加一个开关,你会把开关放在哪里?为什么?

4 所有电路都有的共同元件是什么?

Complex Circuits

There are many different types of electrical components. Each one has a different use in a circuit.
An electrical circuit always has a cell and wires and at least one additional component. Circuits can have more components and these circuits do more than one thing. Look at these more complicated circuit diagrams. Discuss what would happen in each of them.

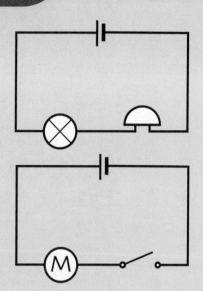

Constructing complex circuits from diagrams

You will need: wires, a cell, a buzzer, a bulb, a motor, a switch...
Work in a small group.

1. Look at the circuit diagrams. Predict what the components will be able to do.
2. Make the circuits shown in the diagrams. Did they do what you expected?

Now make and record some more complex circuit diagrams of your own. Use a cell and wires in every circuit. Each circuit must contain at least two components in addition to the cell and wires. Use the bulb, motor, buzzer and switch (use each up to three times). The circuits must be different from the ones in the diagrams on this page.

3. Make and test your circuits.
4. Record your diagrams.
5. Describe what each circuit does.

复杂的电路

电路元件有很多不同的类型，每种元件在电路中的作用不同。

一个电路总是包含电池、导线，以及至少一个其他元件。电路中可以包含很多元件，而且这些电路可以不止一个功能。观察这些更为复杂的电路图，讨论一下每张图会发生什么。

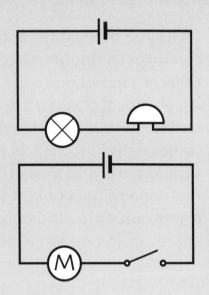

借助电路图搭建复杂的电路

你将需要：若干电线、一节电池、一个蜂鸣器、一个灯泡、一个电动机、一个开关……

和你的小伙伴一起行动。

1 观察电路图。预测这些电路元件能够起到的作用。

2 搭建电路图对应的电路，电路的运行符合你们的预期吗？

现在自己画一些更加复杂的电路图，每个电路中都要使用一节电池和若干导线。每个电路除一节电池和若干导线外至少要包含两个以上的其他元件。请使用灯泡、电动机、蜂鸣器和开关（每一种元件最多使用三次）。要求搭建的电路与本页的电路图不一样。

3 动手制作并测试电路。

4 记录你的电路图。

5 描述一下每个电路的作用。

Finding Faults

Sometimes components in electrical circuits do not work. We say there is a **fault** in the circuit.

Word Box
fault

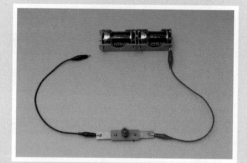

1 The bulb in this circuit is not lit. Why not?

There are many simple things that can stop an electrical circuit working.

Listing faults

You will need: a hand lens, bulb, bulb holder, a cell, wire, a motor...
Work in a small group.

1 Examine the components closely.
2 On sticky notes record possible faults that would stop the circuit working. Using a red pen write a different fault on each sticky note, and with a blue pen record a possible way to fix each fault.
3 Compare your faults with another group.
4 How many faults did you find? Did you miss any that the other group found?
5 Record any you have missed.

Electricians use fault finder guides. These help them when there are electrical problems. The guides have a list of things that can go wrong, and for each fault give ideas about how to fix it.

2 Use the information from your sticky notes to make a fault finder guide.

找出故障

有时，电路中的元件不能正常工作，我们就称电路出现了**故障**。

> **单词表**
> fault 故障

> **1** 电路中的灯泡无法亮起，为什么？

很多小问题都能阻断电路正常运行。

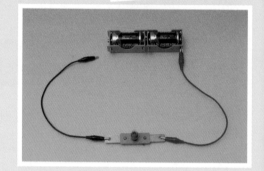

列出故障

你将需要：一个放大镜、灯泡、灯座、一节电池、导线、一个电动机……

和你的小伙伴一起行动。

1 仔细检查（电路中的）元件。
2 将电路可能出现的故障在便利贴上记录下来。在每张便利贴上用红笔写下不同的故障，再用蓝笔写下可能的解决方案。
3 将你们的故障与另一个小组进行对比。
4 你们发现了多少故障？有哪些是其他小组发现而你们没发现的？
5 记录你们没发现的故障。

电气技师们会用到故障排除指南。当遇见电路故障时，故障排除指南能帮助他们解决问题。故障排除指南列出了许多可能会出现问题的地方，也为每种故障列出了解决方案。

> **2** 借助便利贴上的信息来制作一个故障排除指南。

Science Skills

Bulb Brightness – Investigate it!

The brightness of bulbs is not always the same. Have you noticed that in light fittings some bulbs are brighter than others? They are designed that way. It is also possible to change the brightness of a bulb in a simple circuit.

Changing the number of bulbs in a circuit

You will need: two AA 1.5 V cells, a battery holder, six identical bulbs, six bulb holders, seven wires...

You are going to observe what happens to the brightness of the bulb when you add extra bulbs to the circuit.

Read through all the instruction first.

1. Light a bulb in simple circuit. Use the cells, battery holder, a bulb and three wires. The bulb should be bright.

   If the bulb is not lit use a fault finder to help get the bulb to light.

2. What do you think will happen to the brightness of the bulb as you add more bulbs to the circuit? Why?

3. Add the extra bulbs one at a time. You will need an extra wire, bulb and bulb holder for each addition.

4. Record whether the bulbs are brighter, dimmer or the same as the bulb in step 1.

5. Discuss your findings. Why do you think this happens?

> Discuss other ways that the brightness of a bulb in a circuit might be changed.

科学技能

灯泡的亮度——探究一下吧!

灯泡的亮度并不总是一样的,你有没有注意到灯饰中有些灯泡比其他灯泡更亮?就是因为这样的设计,在简单电路中才可以改变灯泡的亮度。

改变电路中灯泡的数量

你将需要:两节1.5伏的五号电池、一个电池底座、六个相同的灯泡、六个灯座、七根导线……

当你在电路中添加额外的灯泡时,观察一下灯泡亮度的变化。首先,阅读下面的实验说明。

1. 在一个简单的电路中,点亮一个灯泡。将用到电池、电池底座、一个灯泡和三根导线。灯泡应该是明亮的。

   如果灯泡没有被点亮,请使用故障排查器,让灯泡重新亮起。

2. 如果你继续在电路中添加灯泡,你认为灯泡的亮度会如何变化?为什么?

3. 每次添加一个灯泡,你都将用到额外的一根导线、灯泡和灯座。

4. 与步骤一的灯泡相比,记录一下你的灯泡是更亮、更暗还是一样亮。

5. 讨论一下你的发现,你认为原因是什么?

讨论一下,是否有其他方法可以改变电路中灯泡的亮度?

Science Skills

Changing Voltage – Investigate it!

The **voltage** is a sort of electric force that makes electricity move round a circuit.
We can think of it as a measure of the push that moves the **electric current**.
Look at these cells.

Word Box
electric current
voltage

1. What do you notice about the voltage?
2. How do you think the size of the voltage changes the brightness of the bulb?

Investigating how changing voltage changes bulb brightness

1. Collect cells like those shown in the pictures above.
2. Which do you think will make the bulb brightest? Why?
3. Which do you think will make the bulb least bright? Why?
4. Place the cells in order. Start with the cell you think will make the bulb brightest on the left to least bright on the right.
5. Discuss how you would investigate whether the size of the voltage changes the brightness of the bulb. Record how you could find out.
6. In a small group carry out the investigation and record your findings.

科学技能

变化的电压——探究一下吧!

电压是某种能让电子在电路中循环流动的力,我们可以把电压看成是能让**电流**发生移动的某种推力。

观察这些电池。

单词表

electric current 电流
voltage 电压

1 关于电压你注意到了什么?
2 你认为电压的大小如何影响灯泡的亮度?

探究变化的电压如何改变灯泡的亮度

1 收集如上图所示的电池。
2 你认为哪个电池能让灯泡最亮?为什么?
3 你认为哪个电池能让灯泡相对最暗?为什么?
4 将电池按顺序排列。从左到右依次为让灯泡最亮的电池到让灯泡相对最暗的电池。
5 讨论一下你将如何探究电压的大小对灯泡亮度产生的影响。记录你是如何探究的。
6 以小组为单位进行实验探究,并记录你们的发现。

Electricity Explained

Scientists often use **models** to help explain difficult concepts. Here are some simple models that can be used to help explain electricity.

We can see, feel and hear the results of electricity, but we cannot actually see what happens inside the wires.

Here are two models that are often used to explain electricity in a circuit. In each model the different parts of the circuit are represented by different things.

Model 1: Water flowing in a pipe

In this model, the cell is represented by a pump. The pump is pumping water around a pipe.

Word Box
model

1 List some ways we see, feel and hear the results of electricity.

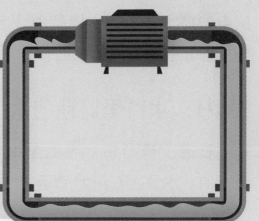

2 What do you think the water represents? What do you think the pipe represents? Complete a table to illustrate this.

Part of circuit	How it is represented
cell	pump

解释电的原理

科学家们通常会运用**模型**来解释复杂的概念。下面的简单模型能用来解释电的原理。

单词表
model 模型

我们能看到、感觉到、听到电所带来的结果，但是我们不能用肉眼观察导线里发生的情况。

📖 **1** 列出一些方式，这些方式是我们能看到、感觉到、听到的电所带来的结果。

下面是两个经常用来解释电路中的电的模型。在每个模型中，电路不同的部分都由不同的东西代替。

模型1：水管里流动的水

在这个模型中，电池是由水泵表示的。水泵正在沿水管运输水。

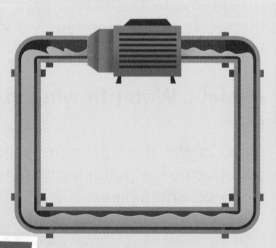

📖 **2** 你认为水代表了什么？你认为水管代表了什么？完成表格来解释一下。

电路元件	表示方式
电池	水泵

Model 2: Passing balls around a circle of people

You will need: a small ball for each person.

Work in a group.

1. Stand in a circle with some of your friends. Each of you should hold a small ball in your left hand.
2. Each person passes their ball to the person next to them in a clockwise direction. Everyone should pass the ball with their left hand.
3. At the same time they should take the ball from the person to their right with their right hand.
4. Everyone should then put the ball into their left hand and pass it on again.

📖 3 Do you think the person saying "pass" in the diagram represents the wire, the electricity or the cell? Why do you think this?

📖 4 Complete a table to illustrate what the different parts represent.

Part of circuit	How it is represented

📖 5 Which of these two models do you think is the best? Explain your choice.

模型2：转圈传球

你将需要：与人数对应的小球。

和你的小伙伴一起行动。

1. 和你的小伙伴一起，围成一个圆圈，每个人左手拿一个小球。
2. 每个人按顺时针方向将球传给相邻的人，每个人都用左手传球。
3. 同时，每个人需要用右手接住右手方向传过来的球。
4. 每个人将球再次放到左手，重复上述步骤。

📕 3 你认为图中说"通过"的人代表的是导线、电流还是电池？为什么？

📕 4 完成表格，来解释不同部分分别代表什么。

电路元件	表示方式

📕 5 你认为这两个模型哪个更好？解释你的选择。

Models for Lighting a Bulb in a Circuit

1. Use a hand lens to look closely inside a light bulb.
2. You will see a thin wire called a **filament**. But what does it do? Why do you think it is there?

Word Box
filament

3. Draw a bulb. Label the filament and explain what you think happens when it is connected in a circuit.

Observing changes to the filament

You will need: two wires, a cell, a bulb holder, a bulb, a switch, a hand lens...

1. Make a circuit to light a bulb. Include a switch.
2. Switch the bulb off then switch on the bulb. What happens to the filament? Why?

Here are two common models used to explain how the bulb is lit.

Model 1: Rope loop model

Look closely at this diagram. It shows children modelling how a bulb is lit in a circuit. This is the rope loop model. Nadia's instructions about how to model this are not very clear:

"The hands holding the string are open so the string can move through them. One person closes their hand a little and another person pulls the string. The hands of the person holding the string more closely feels their hands warm up."

模型：在电路中点亮灯泡

1 使用一个放大镜，仔细观察灯泡的内部。

2 你会观察到名为**灯丝**的线圈。灯丝有哪些作用？你认为灯泡内为什么会有灯丝？

单词表
filament 灯丝

3 画出一个灯泡，把灯丝标注出来。解释一下你认为灯泡接入电路中，灯丝会有哪些改变。

观察灯丝的变化

你将需要：两根导线、一节电池、一个灯座、一个灯泡、一个开关、一个放大镜……

1 搭建电路来点亮灯泡，电路中包含开关。

2 将开关断开，灯泡就不亮了；再将开关闭合，灯泡又亮了。灯丝会发生什么变化？为什么？

下面是两个常见的模型，它们是用来解释灯泡是如何被点亮的。

模型1：绳循环模型

仔细观察这张图，图中展示了孩子们模拟灯泡如何在电路中被点亮，这个模型是绳循环模型。纳迪娅写了一些如何进行模拟的说明，但不是很清楚：

"手持细绳，但是手要张开，这样才能让细绳在手中移动。一个人稍微抓紧绳子，另外一个人拉动绳子。抓紧绳子的人会真切地感到双手发热。"

4 Record a clearer set of instructions. Use Nadia's ideas to help.

5 Why do their hands warm up? What component in the circuit is this model trying to explain?

Performing the rope loop model

You will need: a long piece of string.

1 Work in group and try the rope loop model.

Model 2: Little cup of raisins model

This model is like the ball pass model, but instead of passing a ball you put three or four raisins in small cups. People stand in a circle passing the cups of raisins around the circle when signalled to do so.

One person acts as the bulb and eats a raisin from each cup when it reaches them. The bulb is lit when the cup has raisins to offer.

6 What do you think will happen to the bulb when the raisins run out?

7 What models each part of the circuit in these two models of lighting a bulb?

8 Research other models that explain how electricity works and record one.

4 写下更清楚的一组说明，可以借鉴纳迪娅的想法。

5 为什么他们的手会发热？这个模型在解释电路中哪个元件的原理？

演示绳循环模型

你将需要：一根长绳。

1 和你的小伙伴一起，尝试模拟绳循环模型。

模型2：小杯葡萄干模型

这个模型与转圈传球模型类似，但这里传递的不是球，而是传递装了三颗或四颗葡萄干的小杯子。

人们围成一个圈，当收到指令时，将小杯子沿圆圈传递。

一个人代表灯泡，当杯子传给他时，他就吃掉一颗葡萄干，当杯子里还有葡萄干时，灯泡就可以被点亮。

6 当葡萄干一个都不剩时，你认为灯泡会发生什么？

7 在点亮灯泡的这两个模型中，每一个部分分别代表电路中的哪个元件？

8 研究一下其他解释电路原理的模型，将其中一个记录下来。

Brightness and Buzz Volume

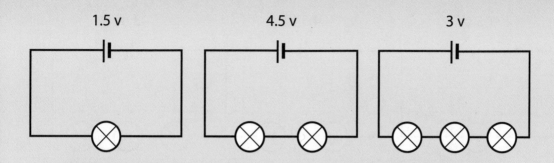

1. Predict the brightness of bulbs in each of these circuits and order them from brightest to dimmest. Work in a group and explain your thinking.
2. What would happen if the bulb was replaced with a buzzer in these circuits? Explain which circuit would have the loudest buzz, and which would have the quietest. Why does this happen?

3. Draw and rank three different circuits of your own that would have bulbs with different brightness.
4. What would happen if the bulb was replaced with a buzzer in your circuits?
5. How do you think the sound of the buzzer would differ in each circuit?

Replacing the bulbs with a buzzer

You will need: wires, bulbs, 1.5 V and 4.5 V cells, a buzzer...

Work in a small group.

1. Construct the circuits shown in the diagrams above, replacing the bulb with a buzzer in each one.
2. Were your predictions about the volume of the buzzer correct?

（灯泡的）亮度和蜂鸣器的音量

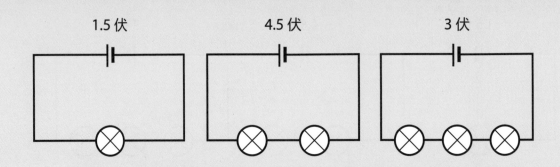

1 请预测上面三个电路中每一个灯泡的亮度，将它们按照从亮到暗的顺序排列。和你的小伙伴一起，交流一下并解释你的想法。

2 如果上图中的灯泡换成蜂鸣器会发生什么？解释一下哪个电路中蜂鸣器的音量最大？哪个电路中蜂鸣器的音量最小？

为什么？

3 自己画出三个不同的电路让灯泡发出不同的亮度，并根据亮度进行排序。

4 如果电路中的灯泡换成蜂鸣器会发生什么？

5 你认为每个电路中蜂鸣器的声音会有哪些不同？

将灯泡换成蜂鸣器

你将需要：导线、灯泡、1.5伏和4.5伏的电池、一个蜂鸣器……

和你的小伙伴一起行动。

1 搭建如上图所示的电路，将每个电路中的灯泡用蜂鸣器替代。

2 关于蜂鸣器的音量，你的预测是否正确？

6 Look closely at these two pictures of circuits. What is the difference between them?
7 The buzzer only works in one of the circuits. Why do you think this happens?

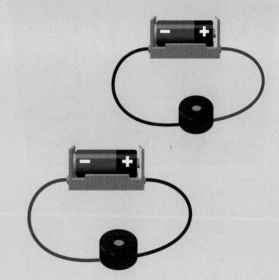

Quietest buzzer challenge

You will need: a buzzer, a 3V cell, a switch, and other electrical components including bulbs, wires, motors...

1 Work in a small group and compete with other groups in the class to make a buzzer sound as quietly as possible. You must still be able to hear the buzz. Include a switch in your circuit so you can turn the buzzer on and off.

2 Draw a circuit diagram showing how you made the volume of the buzzer very quiet.

3 Explain why you chose to make the circuit in the way you did.

4 Compare your buzzer volume with other groups.

5 Record the circuit diagram for another group that has made the buzzer volume very low in a different way.

6 仔细观察两幅图片中的电路，它们有哪些不同？

7 蜂鸣器只在其中一个电路中能正常鸣叫，你认为这是为什么？

音量最小的蜂鸣器挑战

你将需要：一个蜂鸣器、一节3伏的电池、一个开关、灯泡、导线、电动机等其他电路元件……

1 和你的小伙伴一起，与其他小组比赛，搭建电路让蜂鸣器发出尽可能小的声音。在电路中接入开关，这样你可以将蜂鸣器打开或关闭。

2 画一个电路图，展示你们是如何让蜂鸣器的音量变得很小的。

3 请解释你们为什么要这样设计电路。

4 与其他小组比较蜂鸣器的音量大小。

5 如果其他小组用了不同的方式使蜂鸣器的音量变得很小，请记录他们的电路图。

Science Skills

Wires – Investigate it!

Look closely at the wires in this picture. They are not all the same.

1. What are the similarities and differences between them?

Xavier thinks that changing the types of wires in a circuit will alter the brightness of a bulb.

2. Do you agree with him? Explain why to a classmate. Record your ideas.

Questions relating to changing wires

You will need: a selection of different types of wire of different lengths and thicknesses...

1. Work in a small group and look closely at the samples of different wires. Xavier's group chose to investigate changing the length of the wire.

    Their question is "How does changing the length of the wire affect the brightness of the bulb?"

2. List other properties of the wire that could change.

3. What could Xavier's group measure or observe? Use your ideas to write some other questions for Xavier's group.

4. Xavier says they will have to simply observe the brightness of the bulb as there is no way to measure brightness. Is he correct?

科学技能

导线——探究一下吧!

仔细观察图片中的导线。它们不是一模一样的。

1 它们之间有什么异同?

泽维尔认为改变电路中导线的类型能改变灯泡的亮度。

2 你同意他的观点吗?向你的同学解释一下,记录下你的想法。

关于改变导线的问题

你将需要:长度不同、粗细不同的各种类型的导线……

1 和你的小伙伴一起,仔细观察不同类型的导线的样本。泽维尔的小组决定探究导线长度的变化。

他们提出的问题是"导线长度的变化如何影响灯泡的亮度?"

2 列出我们可以改变的导线的其他特性。

3 泽维尔的小组能测量或观察到什么?请思考一下,为泽维尔的小组写下一些其他问题。

4 泽维尔说他们只能简单地观察灯泡的亮度变化,因为没有办法测量灯泡的亮度。他的观点是正确的吗?

The equipment in this diagram is called a data logger.

> 3 What do you think it is used for?
> 4 Do you think Xavier could have used it to help him in his investigation? How?

Xavier's group did not use a data logger. They just looked closely at the brightness of the bulbs. When they carried out the investigation they did not find any change to the brightness of the bulb when they changed the length of the wire.

You are going to try the same investigation. Check to see if your results agree with Xavier's.

If you have a data logger, measure the brightness of the bulb. If not you can observe brightness and make a comparison.

is the brightness of the bulb affected by the length of the wire?

You will need: a bulb, a cell, fuse wire, data logger (if available)...

1 Work in a small group and discuss how you will carry out the investigation. Record your plan.

2 Carry out the investigation and record your findings.

> 5 Did your results match Xavier's? Why is this?
> 6 Did your results match other groups? Why is this?

图中的设备叫作数据记录仪。

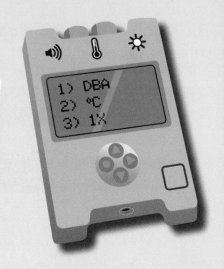

3 你认为数据记录仪是用来做什么的?

4 你认为泽维尔能用数据记录仪帮助他进行实验吗?如果可以,应该如何使用?

泽维尔的小组没有使用数据记录仪。他们只是仔细地观察了灯泡的亮度。

在实验中,当他们改变导线的长度时,并没有观察到灯泡亮度的变化。

你将进行同一个实验,观察你的实验结果是否与泽维尔的一致。

如果你有数据记录仪,请测量灯泡的亮度。如果没有,请观察灯泡亮度并进行比较。

灯泡的亮度受导线长度的影响吗?

你将需要:一个灯泡、一节电池、保险丝、数据记录仪(如果有)……

1 和你的小伙伴一起,讨论一下,你们会如何进行实验并记录你们的计划。

2 进行实验并记录你们的实验结果。

5 你的实验结果与泽维尔的实验结果一致吗?为什么?

6 你的实验结果与其他小组的实验结果一致吗?为什么?

More about Wires

Scientists do not only repeat investigations carried out by other scientists they also investigate further by asking questions of their own.

Investigating wires

You are going to work in a group to plan and carry out an investigation of your own.

1. Look back at the list of questions you asked about wires and choose the one you are going to investigate.
2. Make a full record of how you will do this investigation.

The list of hints below may help you plan it.

Hints:

Record the question.

Predict what you think will happen and why.

List the equipment.

Be clear about variables.

Clearly record what you will do.

Report your findings.

Explain what these findings show.

Note anything you might improve.

了解更多关于导线的知识

科学家们不仅会重复进行其他科学家进行过的实验，而且会自己提出更多问题。

探究导线

你将和你的小伙伴一起，制订计划并进行一个你们自己的实验。

1. 回顾一下你们列出的关于导线的相关问题，选择其中一个作为即将要探究的问题。
2. 全程记录你们的实验过程。

下面的提示也许能帮助你们制定实验计划。

提示：

- 记录问题。
- 预测一下你认为会发生什么以及为什么发生。
- 列出实验器材。
- 明确变量。
- 清楚地记录你们的实验步骤。
- 记录实验结果。
- 解释这些实验结果说明了什么。
- 记录实验过程中可以改进的地方。

Inventors Linked to Electricity

If we look back at history there are many inventors who have invented useful things that use electricity. The three inventors listed here are very famous.

Alessandro Volta was born in 1745 in Como in Italy. He invented the first battery called the "Voltaic pile".

1. What units do we use to measure cell strength?
2. Why do the units have this name?

Michael Faraday was born in 1791. He lived in London in England. He invented the dynamo, which converted motion into electricity. Many scientists did not believe Faraday's ideas because he did not have a university education – he was completely self-taught.

3. What would you invent that runs on electricity?
4. Choose one of these inventors to research further with a partner. Carry out some research and then make a poster about your chosen inventor and their inventions.

Thomas Edison was born in 1847. He lived in Ohio and New Jersey in America. He invented many things including an improved light bulb, a sound recorder, a movie camera and a microphone.

与电有关的发明家

回顾历史，我们可以看到很多发明家发明了与电相关的且实用的东西。下面是三位十分著名的发明家。

1745年，亚历山德罗·沃尔塔出生于意大利罗马。他发明了世界上第一块电池，叫作"伏打电池"。

1. 我们用什么单位来衡量电池的性能？
2. 为什么单位是这个名字？

迈克尔·法拉第出生于1791年。他居住在英国伦敦。他发明了电动机，电动机能够将动能转换为电能。当时，很多科学家都不认可法拉第的科学观点，因为他没有上过大学，完全靠自学。

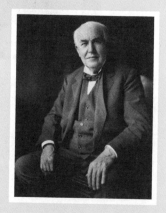

3. 你想发明出哪些需要用电的东西？
4. 从三位发明家中选择一位，和你的小伙伴一起，进行深入调查。做一些研究并制作一张关于这位科学家及其发明的海报。

托马斯·爱迪生出生于1847年。他居住在美国俄亥俄州和新泽西州。爱迪生发明了很多东西，包括改良的灯泡、录音机、摄影机和麦克风。

Testing Our Knowledge!

We can use circuits to make a quiz game.

You will need: a buzzer, switch, wires, cell, cell holder...

1. Working in a group make a simple circuit that switches a buzzer on and off.

You are going to use your circuit and your knowledge about circuits to make quiz cards for a quiz game.

For the game you will need:

your circuit, and for each member of your group: four 12 cm wide by 8 cm high rectangular cards, eight paper clips, six 14 cm long wires with plastic coating (with the ends stripped)...

2. Each of you should list five questions about electricity. For each question provide three possible answers. Make sure there is only one correct answer.

3. Compare your questions with those others in your group have written, and each of you choose two different questions each.

4. Make two cards (like the example below) using your own questions.

Q: What happens if we connect a buzzer to the wrong terminals of a cell in a complete loop circuit?

A: it buzzes.
it whistles.
it doesn't buzz.

知识小测验！

我们可以用电路设计一个知识测验游戏。

你将需要：一个蜂鸣器、开关、导线、电池、电池底座……

1 和你的小伙伴一起，搭建一个能控制蜂鸣器打开和关闭的简单电路。

你将用到你的电路和你学到的电路知识制作测验游戏的卡片。

测验游戏中你将需要：你的电路、四张宽12厘米长8厘米的长方形卡片（每个人都有一张）、八枚回形针、六根长14厘米的带塑料绝缘层的导线（两端裸露）……

2 每个人列出五个关于电的问题，每个问题对应给出三个可能的答案，确保正确答案只有一个。

3 将你的问题同小组成员提出的其他问题进行比较，每人选择两个不同的问题。

4 将你的问题制成两张卡片（如下例所示）。

问题：
如果在闭合回路中我们将蜂鸣器的两端与电池两极连接错误，会发生什么？

答案：
蜂鸣器可以发出蜂鸣声。
蜂鸣器可以发出哨声。
蜂鸣器不会发出蜂鸣声。

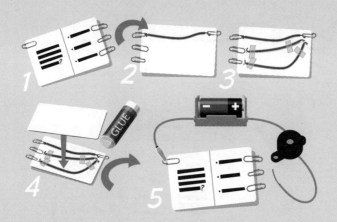

How to make the quiz cards

1. Attach four paper clips to one of the cards, one next to the question and one next to each answer.
2. On the back of the card, attach one end of the stripped-ended wire to the paperclip next to the question and the other end to the paperclip next to the correct answer. Tape the wire to the card.
3. Tape the other two wires to the card. They must not be touching the other answers but should look like they are. Glue another piece of card to the first piece so the wiring can't be seen.
4. Test the card to check that it works in a simple circuit including a buzzer, by attaching the wires of the right answer into the circuit. When the answer is correct the buzzer sounds.
5. Now swap your quiz cards with those of another group. Discuss how you could use these cards in a game to test your knowledge about electricity.
6. Write out the instructions for your game.
7. Play the game to check it works. Make any necessary changes to the instructions. Swap games with another group. Take the quiz.

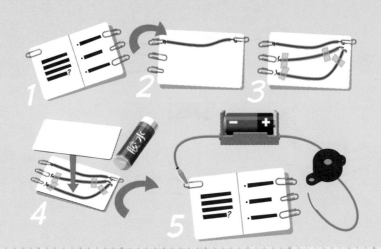

如何制作测验卡片

1 将四枚回形针别在一张卡片上，一枚在问题旁边，另外三枚在三个答案旁边。

2 在卡片的背面，将导线裸露的一端连接到问题旁边的回形针上，另一端连接到正确答案旁边的回形针上，用胶带将导线粘在卡片背面。

3 将另外两根导线也粘在卡片背面，这两根不接触回形针，但看上去应该像接触的样子。再用一张卡片粘在这张卡片上，遮住导线。

4 将连上正确答案的导线接入一个带蜂鸣器的简单电路中。答案正确，蜂鸣器就会发出蜂鸣声。

5 现在将制作好的卡片同另一个小组交换。讨论一下你们要如何在游戏中运用这些卡片，进一步来测试你们关于电的知识。

6 为你们的游戏写下游戏说明。

7 试着玩一玩游戏，检测一下是否成功。如果有必要，对游戏说明可以做出改动。将你们的游戏同另一个小组交换，做一做他们设计的小测验。

Circuit Confusion

Work in a group of five.

1. Draw the circuit diagram symbols for a bulb, a buzzer, a motor, a switch and a cell on stickers. Draw a different circuit diagram symbol on each sticker.
2. Choose one electrical component each and stick it on the front of your clothes.
3. The stickers indicate which component you are. Your arms are wires. Now model making circuits that work.

Look at the three children. One of them has a bulb sticker on their back, the other has a switch and the third has a cell. Their arms are wires.

1. If you made this circuit would it work?
2. What would it do?

Look at the three children again.

3. If you made this circuit would it work?
4. What would it do?

See how many different working circuits you can model that use three or four stickers.

4. Record which stickers are involved in each circuit. Record what each circuit would do if you made it.
5. If a circuit involved all five stickers what would the circuit do?

电路谜团

以五人小组为单位进行活动。

1. 在便利贴上分别画出灯泡、蜂鸣器、电动机、开关和电池的电路图符号。
2. 选择一个电路图元件并贴在你的衣服前面。
3. 便利贴上的符号显示了你代表什么元件,你的手臂是导线。现在连接一个可以正常工作的电路。

观察这三位小朋友。每位小朋友背后都贴着便利贴,其中一位小朋友的背后是灯泡,另一位小朋友的背后是开关,最后一位小朋友的背后是电池。他们的手臂是导线。

1 如果你制作了这个电路,电路会正常工作吗?
2 电路会发生什么?

再次观察这三位小朋友。

3 如果你制作了这个电路,电路会正常工作吗?
4 电路会发生什么?

看看你能用三张或四张便利贴设计多少个不同的且能正常工作的电路。

📓 4 记录每个电路中用到了哪些便利贴,也记录你们搭建的电路有哪些作用。

📓 5 如果一个电路包含了五张便利贴,那么这个电路有什么作用?

Useless Circuits

Look closely at this circuit diagram.

> 1 The bulb will not light up. Why do you think this is?

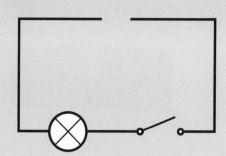

Look closely at this circuit diagram.

> 2 The buzzer will not buzz. Why do you think this is?
>
> 3 Draw some circuit diagrams of your own that will not work and explain why each one does not work.

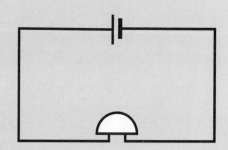

Turning a motor

You will need: a cell, cell holder, two wires, a motor...

The picture shows an electric motor.

1. Work in a small group to make a circuit that includes a motor.
2. Draw the circuit diagram for this circuit. What happens to the motor?
3. What do you think will happen to the motor if you swap over the cell connections? Record your answer.
4. Swap over the cell connections. Was your prediction correct?

> 4 The buzzer on a scoreboard in a sports ground will not work. List all the things you can think of that might be stopping it working.

无用的电路

仔细观察这个电路图。

> 1 灯泡不会亮起,为什么?

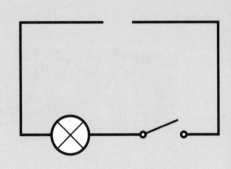

仔细观察这个电路图。

> 2 蜂鸣器不会发出蜂鸣声。为什么?
>
> 📖 3 自己画出一些不能正常工作的电路图,并分别解释原因。

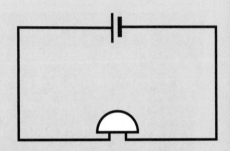

改装电动机

你将需要:一节电池、电池底座、两根导线、一个电动机……

图片展示了一个电动机。

1 和你的小伙伴一起,搭建一个包含电动机的电路。

📖 2 画出搭建好的电路图。电动机会发生什么?

📖 3 如果你将电池的正负极对调,电动机会发生什么?记录你的答案。

4 将电池的正负极对调。你的预测正确吗?

> 📖 4 一个体育场记分牌上的蜂鸣器无法正常工作,请列出你能想到的所有可能的原因。

Changing Circuits

1 List the reasons why a torch may not work.

2 Draw and label the electrical components needed to make a torch work.

Part 2 科学虫大闯关

Circuit Diagram Symbols

Look at some examples of circuit diagram symbols for different components of electrical circuits.

1 Which do you think are good symbols?

2 Which symbols do you think could be improved? Explain why.

3 Choose one of the symbols you think could be improved. Draw it. Then draw your symbol. Explain why you think yours is better.

Circuit diagram symbol	My symbol	I think my symbol is better because:

4 Unfortunately we cannot change the symbols. Why is this?

Remembering Symbols

1 Draw the circuit diagram symbols.

cell	wire
bulb	buzzer
motor	switch

Complex Circuits

1. **Make and record some complex circuit diagrams.**
- Use a cell and wires in each circuit. Use a bulb, motor, buzzer and switch (use each a maximum of three times).
- Each circuit must contain at least two components in addition to the cell and wires. Think of new circuits that you haven't encountered before.

2. Record your diagrams and describe what each circuit does.

This circuit _____

_____

This circuit _____

_____

This circuit _____
_____

This circuit _____
_____

This circuit _____
_____

Finding Faults

1. Make a simple guide to help find faults when circuits are not working. Here's an example.

Electric circuit fault finder	
Fault	Fix
The bulb is broken.	Replace the bulb.

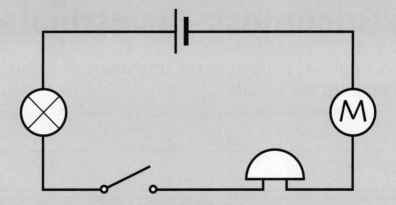

The bulb in this circuit is not working. The motor is working.
2 **List the possible reasons why the bulb is not working.**

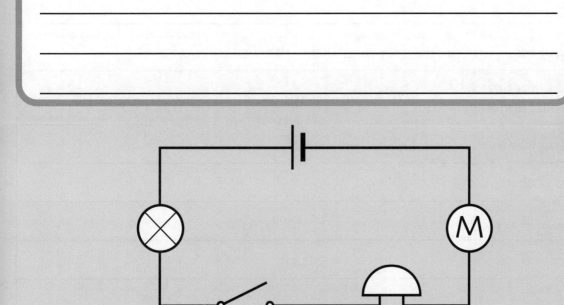

The bulb and motor in this circuit are not working.
3 **List the possible reasons why the bulb and motor are not working.**

Science Skills

Bulb Brightness – Investigate it!

1. Predict what will happen to the brightness of a bulb as you add more bulbs to the circuit.

2. Why do you think this will happen?

Now test your prediction by building the circuits.

Number of bulbs added	Bulb brightness compared to previous brightness
1	
2	
3	
4	
5	

3. What happened to the brightness of the bulb? Was your prediction correct?

4. Why do you think this happens?

Science Skills

Changing Voltage – Investigate it!

1. Draw pictures of three different cells in order of their voltage size. Start with the cell you think will make the bulb brightest. Finish with the one that will make it least bright.

brightest bulb ⟶ least bright bulb

2. How does the size of the voltage change the brightness of the bulb? Record how you could find out.

3. In a small group complete the investigation by connecting cells of different voltage into a circuit with a bulb. Record your findings.

Electricity Explained

1. List six examples of where we see, feel or hear the results of electricity. Identify whether we see, feel or hear it.

Example	Type of effect (see/feel/hear)

Model 1: Water flowing in a pipe

2. What do you think the water represents? What do you think the pipe represents? Complete the table to illustrate this.

Part of circuit	How it is represented
cell	pump
wire	
electricity flowing/electric current	

3a Look at the picture. Do you think the person saying "pass" represents the wire, the electricity or the cell?

3b Why?

Model 2: Passing balls around a circle of people

4 Complete a table to illustrate what the different parts represent.

Part of circuit	How it is represented
cell	
wire	
electricity flowing/electric current	

5 Which of these two models do you think best explains electricity? Explain your choice.

Models for Lighting a Bulb in a Circuit

1 Draw what you can see inside a bulb.

2 Label the filament.

3 What happens to the filament when the bulb is switched on in a circuit?

_____

4 Why does this happen?

_____

5 Write a set of instructions to explain how to carry out the rope loop model.

6 What models each part of the circuit? Complete the tables.

The rope loop	
Part of circuit	How it is represented
cell	
wire	
filament	
electric current	

Little cups of raisins	
Part of circuit	How it is represented
cell	
wire	
filament	
Electric current	

7 Think of and record another model that explains how electricity works.

Brightness and Buzz Volume

1 Draw three different circuits that would have bulbs with different brightness. Draw the brightest on the left and the least bright on the right.

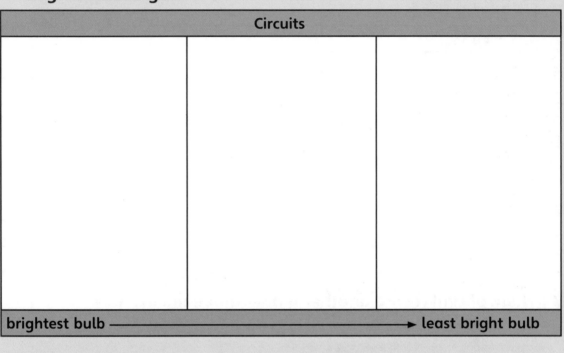

brightest bulb ⟶ least bright bulb

2 What would happen if the bulb was replaced with a buzzer in these circuits? Explain which circuit would have the loudest buzz, and which would have the quietest. Why does this happen?

3 Record a circuit diagram showing how you could make the volume of the buzzer very quiet.

4 Explain why you chose to make the circuit in the way you did.

___

___

___

___

5 Record the circuit diagram for another group that has made the buzzer volume very low in a different way.

6 Why did this work?

___

___

Science Skills

Wires – Investigate it!

Xavier thinks that changing the types of wires in a circuit will alter the brightness of a bulb.

1. Explain why you agree or disagree with Xavier.

2. List the equipment you would need to test whether changing the length of the wire in a circuit affects the brightness of the bulb.

3a What other questions could you ask about how different wires affect the brightness of the bulb? Think of three other questions. List these in the table under Xavier's question.

3b Identify what you could change and measure or observe to try to answer each question. Record these next to the question.

Question	Change	Measure/Observe
How does changing the length of the wire affect the brightness of the bulb?		

4 Is the brightness of the bulb affected by the length of the wire? Record your plan of how to carry out this investigation.

5 Record your results.

6 Did your results match Xavier's?

7 Explain your results. Why does this happen?

8 Did your results match other groups? Why is this?

More about Wires

Look back at the questions you asked about how different wires affect the brightness of the bulb.
Choose the one you are going to investigate and make a full record of how you will carry out this investigation.

The question I am going to investigate is

_____
_____

Prediction

I think that _____

because _____
_____

Equipment I will need to carry out the investigation

_____
_____
_____

We will change _____

We will measure/observe/compare _____
_____

To make it a fair test we will keep these variables the same ____
_____

Write out your method.

Record your results.

Our results show

We could improve what we did by

Testing Our Knowledge!

1. List five questions (Q) about electricity. For each question provide three possible answers (A). Only one of them should be the correct answer.

You can use reference materials to help.

Q1 _____

A1 _____

A2 _____

A3 _____

Q2 _____

A1 _____

A2 _____

A3 _____

Q3 _____

A1 _____

A2 _____

A3 _____

Q4 _____

A1 _____

A2 _____

A3 _____

Q5 _____

A1 _____

A2 _____

A3 _____

These questions and answers can now be used to make quiz cards for a game to play with classmates.

2   Write out the instructions for your game.

Circuit Confusion

1. How many different working circuits can you think of that use three or four electrical components?
2. Record what each circuit would do if you made it.
3. Also record, in the table below, how many components are involved in each circuit. Which components are they?

Description of working circuit	Number of components	Components involved

4. What would a circuit do if it involved all five components?

_____
_____
_____
_____
_____
_____
_____

Useless Circuits

1a Draw two circuit diagrams of circuits that will not work.
1b Explain why each one does not work.

Circuit diagram	Circuit diagram
The circuit will not work because _____	The circuit will not work because _____

2 Draw the circuit diagram for a circuit that includes a motor and a cell.

3 What will happen to the motor if you swap over the cell connections?

4 The buzzer on a scoreboard in a sports ground is not working. List all the things you can think of that might be stopping it working.

Fascinating Facts about Electrical Components

Research some interesting facts about each of these electrical components. Record the facts inside the outline of each component.

Glossary

circuit diagram symbol: symbols used for electrical components

complete circuit: a circuit with a battery (or a cell) in it without a break in the loop

electric current: electricity moving around a complete loop (circuit)

fault (electrical): reason why a component in a circuit is not working

filament (bulb): thin wire inside the bulb that heats up and glows giving light when the bulb is lit

model (scientific): a simple model used to explain something scientifically difficult to understand

series circuit: components in a circuit that are part of the same complete loop

voltage: the push that moves the electric current

词汇表

电路图符号：表示电路元件的符号

闭合电路：包括蓄电池（或电池）且没有损坏的完整电路

电流：在闭合回路（电路）中流动的电子

故障（电的）：造成电路无法正常工作的原因

灯丝（灯泡）：灯泡内部在灯泡点亮时发光、发热的细丝

模型（科学的）：用于解释人们难以理解的事物的简单模型

串联电路：电路元件沿着单一路径连接的闭合回路

电压：能让电流流动的推力